中国水利教育协会
高等学校水利类专业教学指导委员会

U0167043

全国水利行业"十三五"规划教材（普通高等教育）
江苏省高等学校重点教材（编号：2021-2-066）

Pump and Pumping Station
水泵及水泵站（双语教材）

成立　王川　金燕　周济人　编著

中国水利水电出版社
www.waterpub.com.cn
·北京·

内 容 提 要

本书共 10 章，第 1～5 章为水泵篇，第 6 章为泵站水泵机组选型篇，第 7～10 章为泵站工程篇。

本书适用于水利水电工程、农业水利工程、能源与动力工程（水动方向）等专业，也可用于给水排水工程专业，还可供从事水利工程、市政工程等专业的工程技术人员参考。

The book is divided into ten chapters. Chapters 1 to 5 are water pumps. Chapter 6 is the matching part of pump unit selection for pumping station. Chapters 7 to 10 are pumping stations.

This book is applicable to Water Conservancy and Hydropower Engineering，Agricultural Water Conservancy Engineering，Energy and Power Engineering（Water Conservancy and Hydropower Power Engineering）and other majors，as well as Water Supply and Drainage Engineering. This book can also be used as a reference for engineering and technical personnel engaged in water conservancy engineering，municipal engineering，etc.

图书在版编目（CIP）数据

水泵及水泵站 ：双语教材 ＝ Pump and Pumping
Station ：汉、英 / 成立等编著. -- 北京 ：中国水利
水电出版社，2021.12
全国水利行业"十三五"规划教材（普通高等教育）
江苏省高等学校重点教材
ISBN 978-7-5226-0389-6

Ⅰ．①水… Ⅱ．①成… Ⅲ．①水泵－高等学校－教材
－汉、英②泵站－高等学校－教材－汉、英 Ⅳ．
①TV675

中国版本图书馆CIP数据核字(2022)第009759号

	全国水利行业"十三五"规划教材（普通高等教育） 江苏省高等学校重点教材	
书　　名	**Pump and Pumping Station 水泵及水泵站（双语教材）** Pump and Pumping Station SHUIBENG JI SHUIBENGZHAN (SHUANGYU JIAOCAI)	
作　　者	成立　王川　金燕　周济人　编著	
出版发行	中国水利水电出版社	
	（北京市海淀区玉渊潭南路 1 号 D 座　100038）	
	网址：www.waterpub.com.cn	
	E - mail：sales@waterpub.com.cn	
	电话：(010) 68367658（营销中心）	
经　　售	北京科水图书销售中心（零售）	
	电话：(010) 88383994、63202643、68545874	
	全国各地新华书店和相关出版物销售网点	
排　　版	中国水利水电出版社微机排版中心	
印　　刷	清淞永业（天津）印刷有限公司	
规　　格	184mm×260mm　16 开本　15.25 印张　427 千字	
版　　次	2021 年 12 月第 1 版　2021 年 12 月第 1 次印刷	
印　　数	0001—2000 册	
定　　价	**45.00 元**	

前言

　　"水泵及水泵站"是水利水电工程、农业水利工程（原农田水利工程）专业的一门主要专业课程。在高等学校水利学科教学指导委员会指导下，扬州大学、武汉大学、河海大学等高校相继出版了适合不同学校、地区的大学中文版教材。近年来，随着水利专业国际化的不断深入，迫切需要一本适合学生学习的双语教材，以满足教学需要。本书根据相关专业的教学大纲，并参照其他相近专业的要求进行编写。

　　本书基于编者六年双语教学实践，对以往的教学和教材进行了总结，在内容上进行了精选，加强了基本理论、基本概念和基本方法的阐述，同时重视理论联系实践，并注意反映国内外新的科学技术成果。限于篇幅，对于部分需在研究生阶段进一步深入学习的内容，包括水泵在特殊条件下的运行，泵站水锤计算和防护，水泵机组振动、噪声和故障分析等，可参阅已出版的中文相关教材或参考书。

　　扬州大学成立教授主持编著本书，其他编著者有王川教授、金燕副教授、周济人教授。作者对本书编写过程中扬州大学水利科学与工程学院领导、老师的关心表示感谢。参加本书翻译、校对的有张聪聪、周曼、陈伟、蔡瑞民、张小雨、徐文涛、雷帅浩、刘志泉等同学。

　　本书出版得到了扬州大学出版基金、扬州大学品牌专业、江苏省高等学校重点教材和江苏省优势学科、国家一流本科专业建设点项目资助。

　　由于编者水平有限，书中疏漏在所难免，热忱欢迎读者指正。

编 者

2021 年 8 月

Preface

Pump and Pumping Station is a major professional course of Water Conservancy and Hydropower Engineering, Agricultural Water Conservancy Engineering. Under the guidance of the Steering Committee for Water Conservancy Disciplines in Higher Education Institutions, Yangzhou University, Wuhan University, and Hohai University have successively published teaching materials suitable for different schools and regions. In recent years, with the deepening of the internationalization of the water conservancy profession, a bilingual textbook suitable for students is urgently needed to meet the teaching needs. This book is compiled according to the syllabus of related majors and the requirements of other similar majors.

This book is based on the author's six-year bilingual teaching practice. It summarizes previous teaching and textbooks, selects content and strengthens the explanation of basic theory, basic concepts and basic methods. At the same time, it attaches importance to theory and practice, and pays attention to reflect the new scientific and technological achievements at home and abroad. Due to space limitations, for some content that requires further in-depth study at the graduate level, including the operation of pumps under special conditions, calculation and protection of water hammer in pumping stations, vibration, noise and fault analysis of pump units, etc., you can refer to published Chinese Textbook or reference book.

The author in chief of this book is Professor Cheng Li from Yangzhou University. The other authors of this book is Professor Wang Chuan, Associate Professor Jin Yan and Professor Zhou Jiren of Yangzhou University. The authors expressed his gratitude to the leaders and colleagues of the College of Hydraulic Science and Engineering of Yangzhou University for their concern. Participants in the proofread of this book were Zhang Congcong, Zhou Man, Chen

Wei, Cai Ruimin, Zhang Xiaoyu, Xu Wentao, Lei Shuaihao, Liu Zhiquan and other students.

This book is supported by Publishing Fund of Yangzhou University, Yangzhou University Brand Major, Key Teaching Textbook of Jiangsu Higher Education Institutions, A Project Funded by the Priority Academic Program Development of Jiangsu Higher Education Institutions (PAPD) and Project of National First - class Undergraduate Major.

Due to the limitations of the author, omissions in this book are inevitable, and readers are warmly welcome to correct them.

Authors
August, 2021

目录

Contents

第1章 绪 论

1.1 水泵及水泵站的作用与组成

水泵又称抽水机，是一种转换、传送能量的机械。水泵以叶片式为主，通常由电动机或内燃机驱动，动力机驱动水泵运转，通过叶轮的旋转，动力机的机械能通过水泵转换为水的动能和势能，也就是把动力机的能量传送给水，达到提水和增大水压的目的。改革开放以来，我国的水泵制造生产迅速发展，大量的水泵在工业、农业及国民经济其他各部门投入应用，为推动经济建设和社会发展发挥了巨大作用。

水泵站（简称"泵站"）又称抽水站，它是应用动力机驱动水泵进行抽水、增压，并通过一系列建筑物进行输水的工程设施，常被应用于农田灌溉和排涝，跨流域调水，城市给水及排水，工厂、矿山的供水及循环水，抽水蓄能，城市水环境治理等领域。

各类泵站的用途虽然不同，但其组成基本相同，一般包括取水、引水建筑物，前池、进水池（或集水池），拦污清污设备，泵房，主机组及辅助设备，出水建筑物，变电站，交通桥及附属建筑物。

基本概念：

水泵——能量转换的一种机械，它是将动力机的机械能传递给被抽送的水体，从而使水体的能量增加（图 1.1）。

水泵装置——水泵、进出水管路（流道）的总和（图 1.2）。

图 1.1 水泵

Fig. 1.1 Water pump

图 1.2 水泵装置

Fig. 1.2 Water pump system

图 1.3 泵站工程
Fig.1.3 Pumping station project

抽水装置——水泵、动力机、传动设备、管路及其附件的总和。

泵站——抽水装置、泵房、辅助设备、附属设备、电气设备以及为保证机组正常运行时修建的建筑物（泵房）、上下游连接建筑物的总和。

泵站工程——泵站、上下游河道及附属水利工程（闸、涵、渡槽等）的总和（图 1.3）。

1.2 国内外水泵站建设概况

1.2.1 国内泵站发展概况

我国运用水泵提水灌溉排涝始于 20 世纪初。沿太湖的浙江杭州嘉湖地区和江苏苏锡常地区最早采用小型内燃机、蒸汽机带动水泵抽水；最早于 1924 年在江苏武进建成电力泵站。

20 世纪 50 年代中期开始，我国泵站事业一直以很高的速度发展，最早建成规模较大的有：江苏省丹阳珥渎河上电力提水灌区、河北省静海团泊洼排水泵站、山西省夹马口泵站、陕西省渭北高原提水灌区。此外，江苏、浙江、福建等省还建有水轮泵站、水锤泵站等。

20 世纪 60 年代初，以江苏省江都泵站的兴建为标志，我国开始运用大型泵机组于农田排灌和区域调水工程。1961 年江都第一抽水站建成，至 1975 年已建大型电力泵站 4 座，成为江水北调和南水北调东线第一期的源头泵站。随着农业生产的发展，长三角、江汉平原、珠三角、杭嘉湖地区洞庭湖地区及苏北里下河地区等均建有各种类型低扬程大流量泵站，规模较大泵站有：淮安二站、谏壁排灌站、高港泵站、皂河一站、驷马山泵站、临洪东西站、东深供水太园泵站、太浦河泵站、常熟泵站、秦淮新河泵站、黄天湖泵站、沉湖泵站、高潭口泵站、樊口泵站、沙河口泵站、仙桃泵站等。

我国北方地区的泵站主要以高扬程为主，1974 年建成甘肃景泰川提水第一期工程。1984 年开始动工兴建第二期工程，建成 18 个梯级，泵站总净扬程 602m。陕西省抽黄灌溉工程，沿黄河分别在韩城市禹门口、合阳县东雷、潼关县港口三处兴建大型高扬程泵站，其中东雷泵站分 8 个梯级，泵站总净扬程 311m。华北及西南等地江、河、水库沿岸，由于水源水位变化幅度很大，不适宜建固定站，兴建了一批浮动式泵站。其中 1971 年开始兴建的山西省大禹渡站 7 级提水，总扬程 345m，其中一级站在黄河岸边，7 台浮动泵车上装 14 台口径 500mm 双吸离心泵。

为了解决区域水源紧缺问题，60 年代以来我国兴建了 7 个以上跨流域调水工程，大多在东部沿海地区。引滦入津调水工程采用 5 级提水，全线共兴建大型泵站 4 座，安装大型轴流泵 27 台。山东引黄济青调水工程共建大型泵站 5 座，总净扬程为 45m，安装大型轴流泵 30 台。南水北调东线工程干线总长 1156km，沿线通过 13 个梯级泵站逐级提水北上，总扬程 65m，一期已建成，二期工程从 2020 年起进入规划实施阶段，规划线路长度

1785km，新建 25 座泵站，将一期工程的 $500m^3/s$ 扩大到 $870m^3/s$，抽引长江水量从一期 87.7 亿 m^3 提高到 165 亿 m^3。滇中引水工程、珠江三角洲水资源配置等一批重大工程正在逐步实施中。

截至 2019 年，全国已建各类固定泵站超过 50 万座，数量遥居世界各国首位，对于抵御自然灾害，保证农业稳产高产，实现农业机械化、电气化，对促进国民经济持续发展及改善生态环境等发挥了巨大作用。

1.2.2 国外泵站发展概况

世界各国大型泵站的建设期主要集中在 20 世纪 80 年代前，主要用于农田灌溉及排水、跨流域调水、围海造田、开垦干旱土地及沼泽地、防洪防潮、城市给水及排水、抽水蓄能等。

美国于 20 世纪 40 年代末开始在哥伦比亚河建大古力泵站，抽提大古力水库水用于灌溉，安装泵机组 12 台套，其中 1949—1951 年安装立式混流泵 6 台，1973 年后相继增加 6 台套抽水蓄能机组。1970 年开始投入运行的圣路易斯提水工程拥有两座泵站，一座 3 台套离心泵站，另一座 3 台套全调节混流泵站。1957 年开工、1973 年底主要工程竣工的加利福尼亚州调水工程拥有当时世界上提水扬程最高、单机功率最大的泵站——爱德蒙斯顿泵站。该站安装 14 台 4 级立式离心泵，泵站净扬程 587m。20 世纪 60 年代末开始建设的中央亚利桑那工程，从科罗拉多河引水到图森，8 级提水，总扬程 884m，其中第一级哈瓦苏泵站，6 台机组，扬程 251m。科罗拉多河引水工程号称美国七大现代化工程奇迹之一，跨越 390km 的沙漠和山地，总提水扬程 493m，其中包括 5 座泵站。

20 世纪 60 年代末至 70 年代以来，日本陆续兴建了相当数量的大型排水泵站。1971—1973 年新潟市西南的西蒲原地区建成的新川河口排水站，安装 6 台套叶轮直径 4.2m 的灯泡贯流式全调节轴流泵，该站开停机过程全部实现自动化。1975 年建成的三乡排水站是日本最大的混流泵站，水泵采用柴油机驱动。1980 年建成的毛马泵站，安装叶轮直径 4m 立式轴流泵，泵站采用蜗壳进水、蜗壳出水双向流道结构布置型式。1996 年建成的日光川泵站是日本最大叶轮直径（$D=4.6m$）的立式轴流泵站。日本低扬程排水泵站大多采用贯流泵、斜轴泵和潜水电泵，已建叶轮直径 1m 以上贯流泵站 70 余座。

荷兰地势低洼，全国有 $1/3\sim1/2$ 土地在海平面以下。由于大规模围海造田和开垦沼泽地等，排水问题突出，早在 13 世纪便用风车抽水，工业革命以后机电提水泵站事业快速发展。代表性的大型泵站如 1973 年于北海运河入海处兴建的爱茅顿排水泵站，安装大型轴流贯流泵 4 台，泵叶轮直径 3.94m。须德海围垦工程采用斜式安装的轴流泵，单机流量为 $10m^3/s$，而泽顿泵站采用卧式安装的大型轴流泵，泵叶轮直径 3.6m，单机流量 $37.5m^3/s$，扬程仅 1.2m。20 世纪 60 年代荷兰水泵制造商开发了一种新型"混凝土蜗壳泵"，在荷兰圩区排水及灌溉工程中和世界许多地区广泛采用。如在苏玛圩区修建的泵站采用立式混凝土蜗壳混流泵形式，共安装 3 台机组。为便于检修，水泵叶轮以至导叶采用可轴向抽芯的结构。荷兰还广泛使用大直径螺杆泵站进行提水排灌。

苏联水泵设计研究水平较高，建大型泵站较早较多。其中轴流泵站如库拉霍夫和莫斯科运河泵站等。1973 年开始投入运转的卡尔申提灌泵站，将阿姆河水抽送至塔里马让水库，7 级提水，每级各安装 6 台套全调节立式轴流泵，泵站总扬程 156m。

其他国家如印度、埃及等，泵站工程也比较发达。印度伦卡兰萨-贝卡尼尔灌溉工程在拉贾斯坦运河左岸建有 4 座大型泵站；泰维灌溉工程中的提水泵站安装 6 台（套）立式水泵，扬程 30m，单泵流量为 1.7m³/s。埃及于 20 世纪末开始兴建东水西调和西水东调大型水利工程。新河谷工程在纳赛尔湖边的图什卡建一座流量近 300m³/s 的大型泵站。

Chapter 1   Introduction

1.1   Functions and composition of pumps and pumping stations

Water pump is a machine that converts and transmits energy. The water pump is mainly vane type. Generally, the water pump is driven by the motor or internal combustion engine, and the power machine drives the water pump to operate. Through the rotation of the impeller, the mechanical energy of the power machine is converted into the kinetic energy and potential energy of the water through the water pump, that is to say, the energy of the power machine is transmitted to the water supply, so as to achieve the purpose of water lifting and increasing the pressure of the water. Since Economic Reform and Open up, the manufacturing and production of pumps have developed rapidly. A large number of pumps have been put into use in industry, agriculture and other sectors of the national economy, playing a huge role in promoting economic construction and social development.

Pumping station is an engineering facility that uses power machine to drive water pump to pump and pressurize water, and carries out water delivery through a series of buildings. It is often used in farmland irrigation and drainage, inter basin water transfer, urban water supply and drainage, water supply and circulating water for factories and mines, pumped storage, urban water environment treatment and other fields.

Although the purpose of each type of pump station is different, its composition is basically the same, generally including: water intake and diversion structures; forebay, intake pool (or collecting pool); trash retaining and cleaning equipment; pump house; main unit and auxiliary equipment; outlet buildings; transformer station; traffic bridge and auxiliary buildings.

Tips:

Water pump—a kind of machinery for energy conversion, which transfers the mechanical energy of the power machine to the water pumped, so as to increase the energy of the water (Fig. 1.1).

Pump system—the sum of water pump, inlet and outlet water pipelines (flow channels) (Fig. 1.2).

Pumping device—the sum of water pump, power machine, transmission equipment, pipeline and its accessories.

Pumping station—the sum of pumping system, pump house, auxiliary equipment, ancillary equipment, electrical equipment, buildings (pump house) built to ensure the normal operation of the unit, upstream and downstream connecting buildings.

Pumping station project—the sum of pumping station, upstream and downstream river courses and auxiliary water conservancy projects (sluice, culvert, aqueduct, etc.) (Fig. 1. 3).

1. 2   Overview of pumping station construction at home and abroad

1. 2. 1   Development of pumping stations in China

The use of water pump for irrigation and drainage began in the early 20th century. Small internal combustion engines and steam engines ware first used to drive water pumps to pamp water in Jiahu area in Hangzhou, Zhejiang along Taihu Lake and the area of Suzhou, Wuxi and Changzhou, the first electric power pumping station was built in Wujin, Jiangsu Province in 1924.

Since the mid-1950s, China's pumping station industry has been developing at a very high speed. The first large-scale pumping stations were built: the electric water pumping area on Erdu River in Danyang, Jiangsu Province, the drainage pumping station in Tuanpowa, Jinghai, Hebei Province, the Jiamakou pumping station in Shanxi Province, and the water pumping area in Weibei plateau, Shaanxi Province. In addition, Jiangsu, Zhejiang, Fujian and other provinces also have water turbine pumping stations, water hammer pumping stations, etc.

At the beginning of 1960s, with the construction of Jiangdu pumping station in Jiangsu Province as a symbol, large pumping units were used in farmland irrigation and regional water diversion projects in China. The first pumping station in Jiangdu was built in 1961, and four large-scale power pumping stations were built in 1975, which became the source pumping stations of the first phase of the water diversion project from the Yangtze River to the north and the Eastern Route of South-to-North Water Diversion. With the development of agricultural production, various types of low-head and large-flow pumping stations have been built in the Yangtze River Delta, Jianghan Plain, Pearl River Delta, Hangzhou-Jiaxing-Huzhou Plain, Dongting Lake and Lixiahe Area in North Jiangsu. The larger pumping stations are 2nd Huai'an Station, Jianbi Irrigation and Discharge Station, Gaogang Pumping Station, Zao River Station, Simashan Pumping Station, Linhong East-West Station, Dongshen Water Supply Taiyuan Pumping Station, Taipu River Pumping Station, Changshu Pumping Station, Qinhuai New River Pumping Station, Huangtianhu Pump Station, Shenhu Pump Station, Gaotankou Pump Station, Fankou Pump Station, Shahekou Pump Station, Xiantao Pump Station, etc.

The pumping stations in northern China are mainly of high head. The first stage of water extraction project in Jingtaichuan, Gansu Province was built in 1974. The second stage project was started construction in 1984 and 18 cascades were built with a total net head of 602m. In Shaanxi Province, large-scale high-head pumping stations are built along the Yellow River at Yumenkou of Hancheng County, Donglei Port of Heyang County and Tongguan Port, of which Donglei pumping station is divided into 8 cascades with a total net head of 311m. Along rivers and reservoirs in North China and Southwest China, a number of floating pumping stations have been built because the water level of water source varies greatly, which makes it unsuitable to build fixed stations. Among them, Dayudu station of Shanxi Province, which was built since 1971, lifts water at grade 7 with a total lift of 345m, the first grade stands on the bank of the Yellow River, and 14 double-suction centrifugal pumps with a diameter of 500mm are installed on 7 floating pump trucks.

In order to solve the problem of regional water shortage, China has built more than 7 inter-basin water diversion projects since the 1960s, mostly in the eastern coastal areas. The Water Diversion Project from Luanhe River to Tianjin City uses 5-stage water lifting, with 4 large pumping stations and 27 large axial flow pumps installed along the line. Five large-scale pumping stations with a total net head of 45m and 30 large-scale axial flow pumps are installed in the water diversion project of Shandong Province. The trunk line of the Eastern Route of South-to-North Water Diversion Project is 1156km long. Water is pumped up to the North step by step through 13 cascades pumping stations with a total head of 65m. Phase I has been completed and Phase II has entered the planning and implementation stage since 2020. The planned line length is 1785km and 25 new pumping stations are built. The 500m³/s of the Phase I project are expanded to 870m³/s. The volume of pumped water from Phase I 8.87 billion parties is increased to 16.5 billion parties. A number of major projects such as water diversion project in Yunnan and water resources allocation in Pearl River Delta are being implemented step by step.

Up to 2019, over 500,000 fixed pumping stations have been built in China, ranking first in the world in quantity. They play a great role in resisting natural disasters, ensuring stable and high yield of agriculture, realizing agricultural mechanization and electrification, promoting sustainable development of national economy and improving ecological environment.

1. 2. 2　Overview of pumping stations development abroad

The construction period of large pumping stations around the world mainly concentrated before 1980s. They are mainly used for irrigation and drainage of farmland, water diversion across river basins, land reclamation by sea, reclamation of arid land and swamps, flood control and moisture proofing, urban water supply and drainage, pumping and storage, etc.

The United States began to build Grand Coulee Pumping Station on the Columbia

River in the late 1940s. Grand Coulee Reservoir was extracted for irrigation and twelve pumping units were installed, of which six were installed with vertical mixed flow pumps from 1949 to 1951 and six pumped storage units were added after 1973. The St. Louis Pumping Engineering, which began operation in 1970, has two pumping stations, one with three centrifugal pumping stations and the other with three fully regulated mixed-flow pumping stations. Construction began in 1957 and the California State Water Project, completed by the end of 1973, has Edmonston pumping station, the world's largest pumping station with the highest water head and the largest single-unit power. Fourteen four-stage vertical centrifugal pumps are installed in this station, with a net pump head of 587m. The Central Arizona Project, which started constructed in the late 1960s, diverts water from the Colorado River to Tucson, with a total head of 884m, including the first-stage Havasu pumping station with six units and a head of 251m. The Colorado Aqueduct, known as one of the seven miracles of modern engineering in the United States, acrosses 390km of deserts and mountains with a total water head of 493m, including five pumping stations.

From the late 1960s to the 1970s, Japan has successively built a considerable number of large-scale drainage pumping stations. From 1971 to 1973, Shinkawa River Estuary Drainage Station was built in Nishikanbara area, southwest of Niigata. Six sets of bulb tubular fully-regulating axial flow pumps with impeller diameter of 4.2m were installed, and the start-up and shut-down process of the station was fully automated. Misato Drainage Station, built in 1975, is Japan's largest mixed-flow pump station with diesel-driven pumps. Maoma pumping station built in 1980 is equipped with vertical axial flow pump with impeller diameter of 4m. The pumping station is arranged in two-way flow channel structure with spiral case inlet and outlet. Nikkogawa pumping station, built in 1996, is a vertical axial flow pumping station with the largest impeller diameter ($D = 4.6$m) in Japan. In Japan, low-head drainage pumping stations mostly use tubular pumps, oblique shaft pumps and submersible pumps. More than 70 tubular pumping stations with impeller diameters of more than 1m have been built.

The Netherlands is low-lying, with between one-third and one-half of the country's land below sea level. Due to the large-scale reclamation of the sea and marshland, the drainage problem is prominent. As early as the 13th century, water pumped by a windmill, after the industrial revolution, the cause of mechanical and electrical water pumping station has developed rapidly. Representative large-scale pumping stations, such as the Ijmuiden drainage pumping station built at the entrance of North Sea Canal in 1973, installed four large-scale axial-flow through flow pumps with impeller diameter of 3.94m. Zuider Zee Reclamation Project adopts inclined axial-flow pump with single unit flow of 10m^3/s, while Zeeden pumping station adopts horizontal large axial-flow pump with impeller diameter of 3.6m, single unit flow of 37.5m^3/s and head of only 1.2m. In the 1960s, Dutch pump manufacturers developed a new type of concrete volute pump, which is widely used in drainage and irrigation projects in Dutch polder areas and many parts of the world. For example, the pumping station built in Suma Polder adopts the form of vertical concrete volute mixed flow pump, with three sets of units installed in total. In order to facilitate maintenance, the pump impeller and guide vane adopt the structure of axial core pulling. In the Netherlands, large diameter screw pumping stations are also widely used for water lifting, drainage and irrigation.

In the former Soviet Union, the level of pump design and research was relatively high, and the construction of large-scale pumping station was earlier. Among them, axial flow pumping stations such as Kulahov and Moscow canal pumping station, etc. Carl Schen pumping station, which was put into operation in 1973, pumped the Amu Darya to Tarimarjan reservoir, with seven stages of water extraction, each stage equipped with six sets of fully regulated vertical axial flow pumps, with a total head of 156m.

In other countries, such as India and Egypt, the pumping station project is also relatively developed. There are 4 large-scale pumping stations on the left bank of the Rajasthmn Canal in the Loon Karan Sar-Bikaner irrigation project in India; six sets of vertical pumps are installed in the pumping station of Tawi irrigation project, with a head of 30m and a single pump flow of $1.7m^3/s$. At the end of the 20th century, Egypt began to build a large-scale water conservancy project of East-to-West Water Diversion and West-to-East Water Diversion. In the new Valley project, a large pumping station with a flow of nearly $300m^3/s$ built in Tushka near the Lake Nasser.

第2章 水泵类型与结构

2.1 泵的定义与分类

2.1.1 泵的定义

泵是一种流体机械，它把动力机的机械能或其他能源，通过工作体的运动，传给被抽吸的流体，使流体的能量增加，达到提升、输送、增压的目的。如泵抽送的介质是水，就称为水泵。用于农田灌溉和排涝的水泵又称为农用水泵，它占我国泵类产品的50%以上。

2.1.2 水泵的分类

水泵有不同的种类，按工作原理可分为三大类。

(1) 叶片泵。利用叶轮高速旋转，叶片与被抽送水体发生力的相互作用，使水的压能和动能增加，以达到抽送水体的目的。叶片泵按照叶轮对水体的作用原理，又分为离心泵、混流泵和轴流泵三种泵型。叶片泵具有效率高、启动迅速、工作稳定、性能可靠、易调节等优点，用途最为广泛。叶片泵的详细分类参见表2.1。

(2) 容积泵。利用泵密闭工作室容积周期性的变化，对水体产生挤压作用，以增加其压能，达到输送水体的目的。根据工作室容积改变的方式，又可分为往复泵和回转泵两种。容积泵不常用于抽送水体。

(3) 其他类型泵。包括只改变液体位能的泵，如水车、螺旋泵等；利用高速工作水流能量来输送水体的射流泵；利用管道中产生的水锤压力进行提水的水锤泵等。

表 2.1   叶 片 泵 的 分 类

Table 2.1   **Classification of vane pump**

叶片泵 vane pump	离心泵 centrifugal pump	单级单吸离心泵 single-stage single-suction type	卧轴式 horizontal shaft 立轴式 vertical shaft
		单级双吸离心泵 single-stage double-suction type	
		多级离心泵 multi-stage type	
	混流泵 mixed flow pump	蜗壳式混流泵 volute type	卧轴式 horizontal shaft 立轴式 vertical shaft
		导叶式混流泵 guide vane type	
	轴流泵 axial flow pump	轴伸式轴流泵 shaft-extension type	立轴式 vertical shaft 斜轴式 inclined shaft 卧轴式 horizontal shaft
		贯流式轴流泵 tubular type	卧轴式 horizontal shaft

2.2 叶片泵的分类

2.2.1 离心泵

2.2.1.1 离心泵工作原理及分类

离心泵属于高扬程叶片泵,其扬程范围从十几米至数千米,流量相对较小,广泛应用于工业、农业、市政、国防等领域。离心泵是利用叶轮旋转时产生离心力而工作的(图 2.1)。在启动前必须使泵内和进水管中充满液体,当叶轮在泵壳内高速旋转时,液体质点在离心力作用下被甩向叶轮外缘,并汇集到泵壳内,使液体获得动能和压能,并沿着出水管输送出去。

离心泵品种规格和结构型式很多,按叶轮进水方式可分为单面进水悬臂式离心泵(简称单吸离心泵)(图 2.2)及双面进水离心泵(简称双吸离心泵)(图 2.3)两种。按泵内叶轮数目可分为单级泵和多级泵两种。

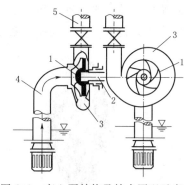

图 2.1 离心泵结构及抽水原理示意

Fig. 2.1 Structure of centrifugal pump and schematic diagram of pumping principle

1—叶轮 impeller;2—泵轴 pump shaft;3—泵 pump;4—进水管 intake pipe;5—出水管 outlet pipe

图 2.2 单吸离心泵

Fig. 2.2 Single-suction centrifugal pump

图 2.3 双吸离心泵

Fig. 2.3 Double-suction centrifugal pump

2.2.1.2 离心泵的构造

离心泵一般由叶轮、吸入室、压出室、泵轴、轴套、轴承、轴封装置、密封环、轴向力平衡装置及底座等组成。

(1)叶轮。又称转轮,是直接与水接触、将机械能传递给水并使水的能量增加的零件。叶轮直接决定水泵性能,因此它是水泵中最重要的部件。

单吸式离心泵的叶轮由前盖板、后盖板、轮毂和叶片四部分组成。离心泵的叶片多为后弯形(弯曲方向与叶轮旋转方向相反),根据是否有前后盖板又可分为封闭式(图2.4)、半封闭式(图 2.5)和开敞式叶轮三种。一般抽清水时采用封闭式叶轮,在污水泵中则多为半封闭式或开敞式叶轮。双吸泵的叶轮相当于两个背靠背的单吸泵叶轮。

（2）吸入室。单级泵吸入室又称泵盖，双级泵吸入室与下部泵壳铸成整体，它的作用是使液体以最小的损失均匀地进入叶轮。

（3）压出室。压出室外壳又称泵壳，其主要作用是以最小的损失汇集由叶轮流出的液体，将液体的部分动能转变为压能，并均匀地将液体导向水泵出口或次级叶轮。

水泵的叶轮、吸入室、压出室以及泵的吸入口、吐出口称为泵的过流部件，它的形状、制造工艺是影响水泵性能、效率和寿命的关键因素。

图 2.4　离心泵封闭式叶轮

Fig. 2.4　Closed impeller

图 2.5　离心泵半封闭式叶轮

Fig. 2.5　Semi-closed impeller

（4）泵轴、轴套及轴承。泵轴是将动力机的功率传给叶轮的部件，必须有足够的扭转强度，用优质碳素结构钢制成。为了防止轴的磨损和腐蚀，在轴上装有轴套，轴套磨损锈蚀后可以更换。

泵轴与叶轮用键连接，并用反向止动螺母固紧。

支承转子的轴承有滚动轴承和滑动轴承两种。单吸泵轴承为两个单列向心球轴承；双吸泵轴承有两种，一种为两个单列向心球轴承，另一种为巴氏合金滑动轴承。

（5）轴封装置。轴封装置用来封闭泵轴穿出泵壳处的间隙，防止漏水和进气。轴封装置主要有填料密封、机械密封及浮动环密封三种。中小型水泵常用填料密封，由填料、水封环、填料压盖组成（图 2.6）。填料又称盘根，一般用石棉绳编制，用石蜡浸透后再压

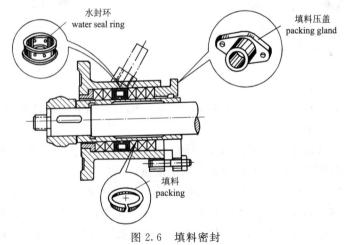

图 2.6　填料密封

Fig. 2.6　Packing seal

成正方形,外涂黑铅粉,填料压紧程度以每秒滴水 1~2 滴为宜。

(6) 密封环,又称减漏环、口环。叶轮进口外缘与泵壳内壁之间的间隙应尽可能小,以减少泵壳中高压水体倒流回泵进口,但在运转时难免发生摩擦而引起叶轮和泵壳的损坏。因此,通常在间隙处的泵壳内安装一道金属环,或在叶轮和泵壳内各安装一道金属环,以便这种环磨损到漏损量太大时予以更换。这种环具有减少漏损和防止磨损两种作用,故又称耐磨环。

(7) 轴向力平衡装置。由于叶轮两侧液体压力不相等而产生轴向力,会使转动部件发生轴向窜动,以致引起磨损、振动和发热,必须采用平衡装置。常见的有平衡孔、平衡管及平衡盘等。

2.2.2 轴流泵

2.2.2.1 轴流泵的工作原理及分类

轴流泵属于低扬程叶片泵,其扬程一般不超过 10m,流量较大。轴流泵是靠叶轮旋转时叶片对水流产生的升力而工作的,与飞机机翼上升力的产生属同一原理。

轴流泵按泵轴安装方式可分为立式、卧式及斜式三种。立式轴流泵应用很广泛,其优点是占地面积小,叶轮浸没在水中,启动方便,电动机安装在水泵上部,不易受潮;但泵轴较长,安装质量要求高。图 2.7 所示为中小型立式轴流泵结构。

卧式轴流泵又分为普通轴伸式、贯流式(又分灯泡式、竖井式)、猫背式、全贯流式(电机泵)等型式。大型卧式轴流泵进出水流道水流平顺,水头损失小,装置效率较高。

斜式轴流泵兼有立式和卧式的优点,某些小型轴流泵立式、卧式、斜式均可安装。

轴流泵按叶片调节方式可分为固定式、半调节式及全调节式三种。固定式是叶片和轮毂体铸在一起,叶片安放角不能进行调节。半调节叶片装在轮毂体上,用螺母压紧,停机后人工转动叶片。全调节轴流泵在运行中或停机后,根据不同的扬程和流量通过一套调节机构来改变叶片的安放角,以适应扬程和流量的需要。大型轴流泵一般为全调节型式。

2.2.2.2 轴流泵的构造

轴流泵由叶轮、导叶、泵壳、轴、轴承、密封、叶片调节机构等零部件组成。

(1) 叶轮。轴流泵叶轮由轮毂体、叶片、导水锥以及轮毂体内的叶片调节机构组成。轴流泵叶轮的叶片一般为 3~6 片,安装在轮毂体上,叶片剖面形状呈流线型,与飞机机翼面相似,故称

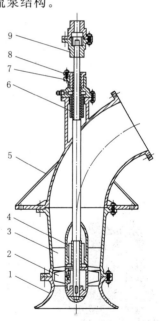

图 2.7 立式轴流泵结构

Fig. 2.7 Structure of vertical axial flow pump
1—喇叭口 bell mouth;2—叶轮 impeller;3—导叶体 guide vane;4—橡胶轴承(下)rubber bearing (lower);5—60°弯管 60° elbow;6—橡胶轴承(上)rubber bearing (upper);7—填料盒座 packing box seat;8—填料压盖 packing gland;9—联轴器 coupling

为翼型。叶片剖面进、出水端顶点连线为翼型的弦，弦与水平面交角称为安放角。固定式叶片的轮毂体为圆柱形，半调节及全调节叶片的轮毂体为球形的一部分，是保证叶片转动任何角度时，能保持叶片与轮毂体相同的间隙。轴流泵叶片呈扭曲形。

（2）导叶。有两个作用：一是扩散水流，回收部分动能；二是将叶轮抽出的水流的旋转运动变为轴向运动。导叶数一般为 6～12 片，导叶体内装有导轴承，起径向支撑作用。

（3）泵壳。用来安装叶轮。轴流泵采用导叶式泵壳，为圆柱桶状，进口端为叶轮室安装叶轮，出口端为出水导叶体，进出水方向在一直线上（图 2.8）。泵壳的作用是汇集从叶轮压出的水流，降低流速，回收部分能量，使水流平顺地流出泵外。

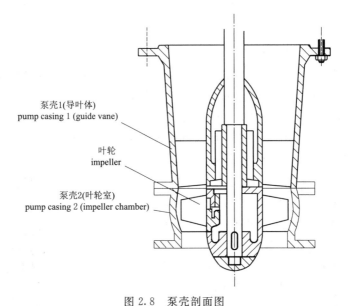

泵壳1(导叶体)
pump casing 1 (guide vane)

叶轮
impeller

泵壳2(叶轮室)
pump casing 2 (impeller chamber)

图 2.8　泵壳剖面图

Fig. 2.8　Pump casing

图 2.9　出水弯管

Fig. 2.9　Outlet elbow

（4）弯管。导叶体后面是弯管（图 2.9），其作用主要是使水流转向，并尽量减少水力损失。

（5）轴与轴承。中小型轴流泵主轴采用优质碳素钢制成并经车削加工，在泵轴穿过轴承及密封部分镀铬，或喷镀不锈钢，或焊接不锈钢套，以防磨损和锈蚀。大型全调节轴流泵主轴是空心的，其中安装叶片调节机构的操纵油管或操纵杆等。

轴流泵采用两种类型轴承：导轴承和推力轴承。

导轴承是用来承受径向力，起径向定位作用。按结构可分为水润滑的橡胶导轴承和油润滑的油导轴承两种，中小型轴流泵大多采用橡胶导轴承，上、下各一个，下导轴承设于导叶体内，上导轴承设于主轴穿过泵壳处。大型轴流泵采用油导轴承。

推力轴承用来承受叶片上轴向水推力及水泵转动部件重量，并维持轴向位置。中小型立式轴流泵采用推力滚珠轴承，大型立式轴流泵的推力轴承装在电动机的上机架上，由推力头、绝缘垫、镜板、推力瓦块和抗重螺栓等组成。轴向力通过电动机主轴推力头、推力瓦块等传递到电动机的上机架。

（6）叶片调节机构。大型轴流泵调节机构通过机械或液压机构来改变叶片的安放角（图2.10），它可以在运行中或停机后不拆卸叶轮的情况下进行调节，故称为全调节。由于叶片可以在一定范围内任意调节，所以水泵高效率区较广。

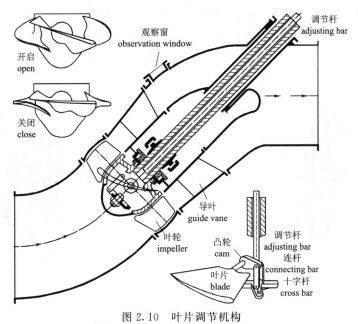

图 2.10 叶片调节机构

Fig. 2.10 Blade adjusting mechanism

2.2.3 混流泵

混流泵又称斜流泵，根据出水室的不同，通常分为蜗壳式和导叶式两种。中小型、低比转速混流泵多为蜗壳式结构，高比转速的混流泵为导叶式结构。

混流泵叶轮（图2.11）的流道较宽，出口边倾斜，其中低比转速的混流泵叶轮与离心泵叶轮相近，高比转速的混流泵叶轮则与轴流泵叶轮相近，其叶片也可以做成可调节式叶片。

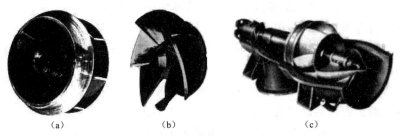

（a）　　　　　　（b）　　　　　　（c）

图 2.11 混流泵叶轮

Fig. 2.11 Mixed flow pump impeller

（a）封闭式；（b）开敞式；（c）半封闭式

(a) Closed type; (b) Open type; (c) Semi-closed type

（1）导叶式混流泵。配有轴向或径向导叶，一般为立式，与蜗壳式泵相比，其泵壳的体积要小些，占地面积也小。如图 2.12 所示，立式导叶式混流泵主要由泵体、泵轴、主轴承和填料盒等组成。与蜗壳式混流泵相比，它的流量范围更大，可达 $1\sim15\text{m}^3/\text{s}$，扬程为 $5\sim20\text{m}$。

（2）蜗壳式混流泵。基本结构与单级单吸离心泵相近。蜗壳式混流泵大多为卧式，但由于混流泵的流量一般比离心泵大，其蜗壳体也较大，为使其支承稳固，将混流泵的泵壳与底座铸为一体，轴承则用螺栓连接在泵体上，靠泵体支承。图 2.13 所示为蜗壳式混流泵，这类泵的流量范围为 $0.02\sim3\text{m}^3/\text{s}$，扬程范围为 $4\sim20\text{m}$。

图 2.12　立式导叶式混流泵

Fig. 2.12　Vertical guide vane mixed flow pump

图 2.13　大型蜗壳式混流泵

Fig. 2.13　Large volute mixed flow pump

2.3　抽 水 装 置

抽水装置由水泵、动力机、传动机构、进出水管（流）道和各种管道附件组成。水泵是整个装置的核心；动力机提供驱动水泵的动力；传动机构用以把动力传递给水泵；管道则是水体从进水池至出水池的过流通道；附件主要起控制水流及辅助调节作用。此外，还有辅助机组设备，如油、气、水系统，电气设备及其控制系统等。这些设备共同完成整个抽水过程。

2.3.1　离心泵抽水装置

图 2.14 所示为离心泵抽水装置示意，各部分作用分述如下。

（1）滤网与底阀。安装在进水管入口，滤网是用来防止水中杂物被吸入泵内，底阀是个单向阀，用来防止水泵启动前充水时的漏失。底阀的水头损失很大，如用真空泵抽气充水，可以取消底阀。

（2）进水管。进水管的布置应避免管中储存有空气，安装时严防进气。

（3）偏心渐扩管。用以扩大进水管径，以减少管道中的水头损失。

（4）真空表。安装在水泵进口处，用以测量水泵进口的真空值。

（5）压力表。安装在水泵出口处，用以测量水泵出口压力。

（6）逆止阀。安装在水泵附近的出水管中，是一个单向阀，在事故停机时自动关闭，阻止水倒流，避免机组高速反转。但是安装逆止阀后，不仅增加水头损失，而且由于它突然关闭,可能产生很大的水锤压力。为了取消逆止阀,可以在出水管口设置拍门或采取其他方法断流;也可采用缓闭逆止阀代替突闭逆止阀。

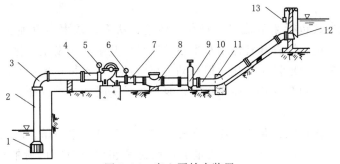

图 2.14　离心泵抽水装置

Fig. 2.14　Pump unit of centrifugal pump

1—滤网与底阀 filter screen and bottom valve;2—进水管 water inlet pipe;3—90°弯头 90° elbow;
4—偏心渐扩管 eccentric diffuser tube;5—真空表 vacuum gauge;6—压力表 pressure gauge;
7—偏心渐缩管 eccentric tapered tube;8—逆止阀 check valve;9—闸阀 gate valve;
10—出水管 outlet pipe;11—45°弯头 45° elbow;12—拍门 flap valve;
13—平衡锤 balance hammer

（7）闸阀。安装在逆止阀后,用以调节水泵流量和功率。

（8）拍门。停机时防止出水池中水倒流入管的一种单向阀门装置。

上述各种设备和附件,并非所有抽水装置都必须具备,应视具体条件取舍。三阀(闸阀、逆止阀及底阀)中的逆止阀和底阀,水头损失很大,能量消耗较大,应予取消。

2.3.2　轴流泵抽水装置

中小型立式轴流泵叶轮安装在进水池水面之下,运行时动力机带动叶轮在水中旋转,水流从喇叭口进入叶轮后,经导叶体、出水弯管和出水管进入出水池。由于叶轮淹没在水下,水泵启动前无须充水,故不需设置底阀。轴流泵不允许关阀启动,出水管上不得装闸阀。为防止停机时水倒流,在出水管出口设置拍门或其他断流装置。

图 2.15 所示为中小型立式轴流泵抽水装置。

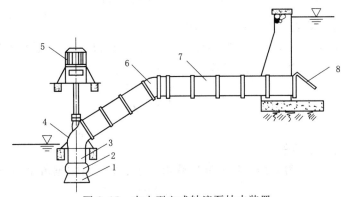

图 2.15　中小型立式轴流泵抽水装置

Fig. 2.15　Small and medium-sized vertical axial flow pump unit

1—喇叭口 bellmouth;2—叶轮 impeller;3—导叶体 guide vane;4—出水弯管 outlet elbow;
5—电动机 motor;6—45°弯头 45° elbow;7—出水管 outlet pipe;8—拍门 flap valve

2.4　水泵的性能参数

水泵的性能参数包括泵的流量、扬程、转速、功率、效率、汽蚀余量（吸上真空高度）等。其中，流量、扬程和转速是基本参数，只要其中一个发生变化，其余参数都会按照一定规律，或多或少地发生变化。但在动力机驱动下，水泵转速一般固定在某一值，故可用其他五个参数的相互变化规律来反映水泵在该转速时的工作性能。

2.4.1　流量

水泵流量分为体积流量和质量流量两种，体积流量是水泵在单位时间内抽送水体的体积，其单位为 m^3/s、L/s 或 m^3/h。质量流量则是泵在单位时间内所抽送的水体的质量，其单位为 kg/s 或 t/h。通常采用体积流量，用符号 Q 表示。

2.4.2　扬程

扬程又称水头，是水泵由叶轮传给单位重量水体的总能量，是能量的概念。可通过泵进、出口断面上的单位总能量 E_1、E_2 的差值表示，其单位以 m 水柱计。

以图 2.16 所示的卧式离心泵为例，以泵轴线为基准面，则

$$E_1 = z_1 + \frac{p_1}{\rho g} + \frac{v_1^2}{2g}, \quad E_2 = z_2 + \frac{p_2}{\rho g} + \frac{v_2^2}{2g}$$

水泵扬程为

$$H = E_2 - E_1 = (z_2 - z_1) + \frac{p_2 - p_1}{\rho g} + \frac{v_2^2 - v_1^2}{2g} \tag{2.1}$$

式中：z_1、z_2 分别为真空表测压点、压力表零位至基准面的垂直距离，低于基准面时取负值，m；p_1、p_2 分别为真空表测压点、压力表零位的绝对压力水头，m；$\frac{v_1^2}{2g}$、$\frac{v_2^2}{2g}$ 分别为泵进、出口断面的流速水头，m。

p_1、p_2 的表读数均为相对值，代入式（2.1）时为

$$H = E_2 - E_1 = (z_2 - z_1) + (H_s + H_d) + \frac{v_2^2 - v_1^2}{2g} \tag{2.2}$$

式中：H_s、H_d 分别为真空表、压力表读数，m。

对于立式轴流泵扬程计算，如叶轮淹没在水中，则 $H_s = 0$，以进水池水面为基准面时，$z_1 = 0$，v_1 很小也可忽略不计，如图 2.17 所示，则

$$H = z_2 + H_d + \frac{v_2^2}{2g} \tag{2.3}$$

2.4.3　功率

水泵的功率有以下两种。

（1）有效功率 N_e。为泵内水体实际所获得的净功率，可根据流量和扬程计算。

$$N_e = \frac{\rho g Q H}{1000} \tag{2.4}$$

（2）轴功率 N_a。水泵在一定流量、扬程下运行时所需的外部功率，即动力机传给水泵轴上的功率，又称输入功率（kW）。轴功率不可能全部传给水体，而是损失一部分功

率后，剩余部分才成为有效功率。

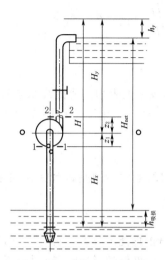

图 2.16 卧式离心泵扬程

Fig. 2.16 Horizontal centrifugal pump head

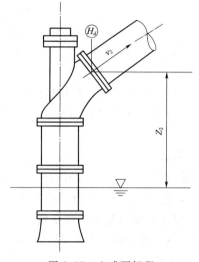

图 2.17 立式泵扬程

Fig. 2.17 Horizontal pump head

2.4.4 效率

有效功率与轴功率的比值为效率 η。

$$\eta = \frac{N_e}{N_a} \times 100\% \tag{2.5}$$

水泵效率反映了水泵传递功率的有效程度，即反映了泵内功率损失的大小，是衡量水泵性能优劣的一项重要技术经济指标。它由泵内水力效率、机械效率及容积效率三个局部效率组成。

（1）机械损失 N_{ml} 与机械效率 η_m。机械损失包括泵轴与轴承的摩擦损失，泵轴与轴封内填料的摩擦损失，叶轮在水中旋转时引起的损失即轮盘损失。水泵克服了机械损失 N_{ml} 之后，把剩下的功率传给抽送的水，这部分功率称为水功率，用 N_h 表示，机械效率 η_m 为

$$\eta_m = \frac{N_a - N_{ml}}{N_a} = \frac{N_h}{N_a} \times 100\% \tag{2.6}$$

其中 $$N_h = \rho g Q_T H_T$$

式中：Q_T 为水泵的理论流量，$\mathrm{m^3/s}$；H_T 为水泵的理论扬程，m。

（2）容积损失与容积效率 η_V。泵内的容积损失指高压侧的水流向低压侧泄漏引起的功率损失。在水泵抽送的水体中，送出泵外的部分流量用 Q 表示，泄漏流量用 q 表示，故有 $Q_T = Q + q$，泄漏水体消耗的能量 $N_q = \rho g q H_T$，令 $N' = N_h - N_q = \rho g Q H_T$，容积效率为

$$\eta_V = \frac{N'}{N_h} = \frac{\rho g Q H_T}{\rho g Q_T H_T} = \frac{Q}{Q_T} \times 100\% \tag{2.7}$$

19

（3）水力效率 η_h。泵内水力损失包括：水泵吸入室、叶槽、流道、出水室等过流部件中的摩擦阻力、漩涡及撞击等引起的水力损失 h，水力效率为

$$\eta_h = \frac{N_e}{N'} = \frac{N'-N_h}{N'} = \frac{\rho g Q H}{\rho g Q H_T} = \frac{H}{H_T} \times 100\% \qquad (2.8)$$

根据式（2.5）～式（2.8）可得

$$\eta = \eta_m \eta_V \eta_h \qquad (2.9)$$

水泵的效率等于该泵的机械效率、容积效率和水力效率的乘积。要提高水泵效率，必须尽量减少机械摩擦和泄漏量，并力求改善过流部件的水力优化设计和提高制造、装配质量。

2.4.5　转速

转速 n 是指叶轮每分钟的转数。水泵铭牌上所标明的额定转速是设计工况时的转速，当转速改变后，水泵工作性能也随之改变。

2.4.6　吸上真空高度和汽蚀余量

吸上真空高度和汽蚀余量均是反映水泵吸水性能或汽蚀性能的参数，它们是确定水泵安装高度和评判水泵发生汽蚀问题的主要参数（有关水泵汽蚀性能的详细内容见本书第 6 章）。

Chapter 2　Types and Structures of Pumps

2.1　Definition and classification of pumps

2.1.1　Definition of pump

Pump is a fluid machinery, which transmits mechanical energy or other energy of power machine to the sucked fluid through the movement of the working body, so as to increase the energy of the fluid and achieve the purpose of lifting, conveying and boosting. If the medium pumped by the pump is water, it is called water pump. Water pumps for irrigation and drainage of farmland, also known as agricultural water pumps, account for a large proportion of pumps in China, exceeding 50%.

2.1.2　Classification of pumps

There are different kinds of pumps, which can be divided into three categories according to their working principle.

(1) Vane pump uses high-speed rotation of impeller and interaction between vane and pumped water body to increase pressure and energy of water, so as to achieve the purpose of pumping water body. According to the working principle of impeller on water body, vane pump is divided into centrifugal pump, mixed flow pump and axial flow pump. Vane pumps have many advantages, such as high efficiency, fast start-up, stable operation, reliable performance, easy adjustment and so on. They are widely used. Table 2.1 shows the detailed classification of vane pumps.

(2) Volumetric pump uses the periodic change of volume in the closed chamber of the pump to squeeze the water body so as to increase its pressure energy and achieve the purpose of conveying water body. It can be divided into reciprocating pump and rotary pump according to the change of workroom volume. Volumetric pumps are not often used to pump water.

(3) Other types of pumps include pumps that only change liquid level energy, such as water trucks, screw pumps, etc. , jet pumps that use high-speed working water flow energy to transport water, and hammer pumps that use water hammer pressure generated in pipes to lift water.

2.2　Classification of vane pump

2.2.1　Centrifugal pump

2.2.1.1　Working principle and classification of centrifugal pump

Centrifugal pump is a high-head vane pump, with a head range from tens to thousands of meters and a relatively small flow rate. It is widely used in industry, agriculture, municipal engineering, national defense and other fields. Centrifugal pumps operate on the principle that centrifugal forces are generated when impellers rotate (Fig. 2. 1). Before starting, the pump and intake pipes must be filled with liquid. When the impeller rotates at high speed in the pump casing, the liquid particles are thrown towards the outer edge of the impeller under centrifugal force and converge into the pump casing, so that the liquid can obtain kinetic and pressure energy and be transported out along the outlet pipe.

There are many kinds of centrifugal pumps, which can be divided into single-side intake cantilever centrifugal pump (single-suction centrifugal pump) (Fig. 2. 2) and double-side intake centrifugal pump (double-suction centrifugal pump) (Fig. 2. 3) according to the inlet mode of impeller. According to the number of impellers in the pump, it can be divided into single-stage pump and multi-stage pump.

2. 2. 1. 2　Construction of centrifugal pump

Centrifugal pump is generally composed of impeller, suction chamber, extrusion chamber, pump shaft, shaft sleeve, bearing, shaft seal device, leakage reducing ring, axial force balance device and base, etc.

(1) The impeller is also called a runner. It is a part that directly contacts water and completes the transfer of mechanical energy to water and increases the energy of water. The impeller directly determines the pump performance and is therefore the most important component of the pump.

The impeller of single suction centrifugal pump consists of front cover plate, rear cover plate, wheel hub and blade. The vanes of centrifugal pump are mostly back-curved (bending direction is opposite to rotation direction of impeller). According to whether having front and rear cover plates, they can be divided into closed impeller (Fig. 2. 4), semi-closed impeller (Fig. 2. 5) and open impeller. Generally, the closed impeller is used for pumping clean water, while the semi-closed or open impeller is used for sewage pump. The impeller of the double suction pump is equivalent to the impeller of the single suction pump with two backs.

(2) The suction chamber of single-stage pump is also called pump cover. The suction chamber of double-stage pump and the lower pump shell are cast as a whole, which is used to make the liquid evenly enter the impeller with the minimum loss.

(3) The main function of the outer casing of the extrusion chamber is to collect the liquid flowing out of the impeller with the smallest loss, change part of the kinetic energy of the liquid into the pressure energy, and guide the liquid evenly to the outlet of the pump or the secondary impeller.

The impeller, suction chamber, extrusion chamber, suction inlet and discharge outlet of the pump are called the overflow parts of the pump. Its shape and manufacturing process are the key factors affecting the performance, efficiency and service life of the pump.

(4) The pump shaft, shaft sleeve and bearing. Pump shaft are components that transmit the power of the power machine to the impeller. They must have sufficient torsion strength and be made of high-quality carbon structural steel. In order to prevent the wear and corrosion of the shaft, the shaft sleeve is installed on the shaft, and the sleeve can be replaced after it is worn and rusted.

The pump shaft and impeller are connected by key, and fixed by reverse stop nut.

There are two kinds of bearings supporting the rotor: rolling bearing and sliding bearing. The bearings of single suction pump are two single row radial ball bearings. There are two kinds of double suction pump bearings, one is two single row radial ball bearings and the other is Babbitt alloy sliding bearing.

(5) The shaft seal device is used to close the gap between the pump shaft and the pump housing to prevent water leakage and air intake. There are mainly three types of shaft seal devices: packing seal, mechanical seal and floating ring seal. Small and medium-sized pumps are usually sealed by packing, water seal ring and packing gland (Fig. 2. 6) . Packing is usually made of asbestos rope, soaked with paraffin, then pressed into a square, coated with black lead powder, and packing degree of compaction is preferably 1 - 2 drops of water per second.

(6) Sealing ring is also called leakage reducing ring and mouth ring. The clearance between the outer edge of the impeller inlet and the inner wall of the pump casing should be small enough to reduce the backflow of high-pressure water in the pump casing back to the pump inlet, but it is inevitable that the impeller and the pump casing would be damaged due to friction during operation. Therefore, a metal ring is usually installed in the pump casing at the clearance, or a metal ring is installed in the impeller and pump casing respectively, so that the ring can be replaced when the leakage is too large. This kind of ring has two functions of reducing leakage and preventing wear, so it is also called wearing ring.

(7) The axial force balancing device produces the axial force due to the unequal liquid pressure on both sides of the impeller, which would cause the rotating parts to move in axial series, resulting in wear, vibration and heating. Therefore, the balancing device must be used. The common ones are balance hole, balance pipe and balance plate, etc.

2. 2. 2　Axial flow pump

2. 2. 2. 1　Working principle and classification of axial flow pump

Axial flow pumps belong to low head vane pumps and their head are generally not more than 10m, and its flowrate is large. The axial-flow pump works by the lift of the blade to the flow when the impeller rotates, which is the same principle as the lift generated by the aircraft wing.

Axial flow pump can be divided into vertical, horizontal and inclined types according to the installation mode of pump shaft. Vertical axial flow pumps are widely used, and their advantages are small floor area, impeller submerged in water, easy to start, motor installed in the upper part of the pump, not susceptible to moisture. But the pump shaft is long and the installation quality is high. Fig. 2.7 shows the structure of small and medium-sized vertical axial flow pump.

Horizontal axial flow pump is divided into ordinary axial extension type, tubular type (including bulb type and shaft type), cat back type, entirely tubular type (motor pump) and other types. The large horizontal axial flow pump has smooth flow in and out of the water channel, small head loss and high efficiency of the device.

Inclined axial flow pump has the advantages of both vertical and horizontal. Some small axial flow pumps can be installed vertically, horizontally and obliquely.

Axial flow pump can be divided into fixed type, semi regulating type and full regulating type according to blade regulating mode. Fixed type is that the blade and hub body are cast together, and the blade setting angle cannot be adjusted. The semi regulating blade is installed on the hub body, pressed with nuts, and turned manually after shutdown. During operation or after shutdown of the fully regulated axial flow pump, a set of regulating mechanism is used to change the setting angle of blades according to different head and flow to meet the needs of head and flow. Large axial flow pumps are generally fully regulated.

2. 2. 2. 2   Construction of axial flow pump

Axial flow pump is composed of impeller, guide vane, pump casing, shaft, bearing, seal, blade adjust mechanism and other parts.

(1) The impeller of axial flow pump consists of hub body, blade, guide cone and blade adjustment mechanism in the hub body. The vanes of the impeller of axial flow pump are generally 3 - 6 pieces, which are installed on the hub body. The profile of the vanes is streamlined and similar to the wing of an airplane. Therefore, they are called wings. The joint line of inlet and outlet ends of blade section is chord of wing, and the angle between chord and horizontal plane is called placement angle. The hub of fixed blade is cylindrical, and the hub of semi-and fully-adjusted blades is part of the spherical shape, which ensures that the clearance between the blade and the hub is the same when the blade rotates at any angle. The vanes of the axial flow pump are twisted.

(2) The guide vane can diffuse the water flow and recover some kinetic energy, and change the rotating motion of the water flow extracted from the impeller into axial motion. The number of guide vanes is generally 6 - 12 pieces. The guide bearing is installed in the guide vane body to provide radial support.

(3) The pump casing is used to mount the impeller. The axial flow pump adopts the guide vane pump shell, which is cylindrical barrel shape. The inlet end is the impeller chamber, and the impeller is installed, the outlet guide vane at the outlet end, and a

straight line of water inlet and outlet direction (Fig. 2. 8) . The purpose of the pump casing is to collect the flow of water from the impeller, reduce the flow rate and recover part of the energy so that the water flows smoothly out of the pump.

(4) The elbow is behind the guide vane (Fig. 2. 9) . Its main function is to divert water flow and minimize hydraulic loss.

(5) The spindles of small and medium-sized axial flow pumps are made of high-quality carbon steel and machined by turning. They are chromed or sprayed with stainless steel or welded with stainless steel sleeves to prevent wear and corrosion. The main shaft of a large fully adjustable axial flow pump is hollow, in which the control tubing or lever with the blade adjusting mechanism is installed, etc.

Axial flow pump uses two types of bearings, one is guide bearing and the other is thrust bearing.

Guide bearings are used to withstand radial forces and act as radial positioning. The structure can be divided into two types: water-lubricated rubber guide bearing and oil-lubricated guide bearing. Most small and medium-sized axial flow pumps use rubber guide bearing, with upper and lower guide bearings, the lower guide bearing inside the guide vane body and the upper guide bearing through the pump casing. Oil-guided bearings are used in large scale axial flow pumps.

The thrust bearing is used to bear the axial water thrust on the blade and the weight of the rotating parts of pump, and maintain the axial position. Small and medium-sized vertical axial flow pumps use thrust ball bearings, and large vertical axial flow pumps have thrust bearings installed on the upper rack of the motor, which are composed of thrust heads, insulation pads, mirrors, thrust tiles and weight-resistant bolts. Axial force is transferred to the upper frame of the motor through the thrust head and thrust pad of the motor spindle.

(6) Blade adjusting mechanism. Large axial flow pump adjustment mechanism is a mechanical or hydraulic mechanism to change the blade placement angle, which can adjust in operation or after shutdown without disassembling the impeller, so called full adjustment (Fig. 2. 10) . Because the blades can be adjusted within a certain range, the high efficiency area of the pump is wide.

2. 2. 3   Mixed flow pump

Mixed flow pump, also known as oblique-flow pump, is usually divided into volute type and guide vane type, according to the different water outlet chambers. Small and medium-sized mixed flow pumps with low specific speed is mostly spiral-shell structure, while high specific speed mixed flow pumps with guide vane structure.

The mixer pump impeller is shown in Fig. 2. 11. It can be seen from the diagram that the flow passage is wide and the outlet edge is oblique. The mixed flow pump impeller with low specific speed is similar to the centrifugal pump impeller, and the mixed flow pump impeller with high specific speed is similar to the axial pump impeller. Its blades can also be made into adjustable blades.

(1) The guide vane mixed flow pump is equipped with axial or radial guide vanes, which are generally vertical. Compared with the volute pump, the pump housing has a smaller volume and occupies a smaller area. Fig. 2. 12 shows the structure of a single guide vane mixed flow pump. It is mainly composed of pump body, pump shaft, main bearing and packing box. Compared with the volute mixing pump, it has a larger flow range, up to $1-15\mathrm{m}^3/\mathrm{s}$, and an elevation of $5-20\mathrm{m}$.

(2) The basic structure of the volute mixed flow pump is similar to that of single-stage single-suction centrifugal pump. Most of the volute mixed-flow pumps are horizontal, but because the flow rate of the mixed flow pump is generally larger than that of the centrifugal pump, the volute of the mixed flow pump is also larger. In order to make the mixed-flow pumps support stable, the pump housing and the base of the mixed flow pump are cast into one, while the bearing is bolted on the pump body and supported by the pump body. Fig. 2. 13 shows the structure of a volute mixed flow pump. The flow rate of such pumps ranges from 0. 02 to $3\mathrm{m}^3/\mathrm{s}$ and the head range from 4 to 20m.

2. 3   Pumping device

Pumping device consists of water pump, power machine, driving mechanism, inlet and outlet pipes (passage) and various pipe accessories. The pump is the core of the whole unit; the power machine provides the power to drive the pump; the driving mechanism transmits the power to the pump; the pipeline is the flow passage of water from the intake pool to the outlet pool; and the accessories are mainly used to control the water flow and auxiliary regulation. In addition, there is auxiliary unit equipment, such as oil, gas and water system, electrical equipment and its control system.

These devices work together to complete the entire pumping process.

2. 3. 1   Centrifugal pump pumping device

Fig. 2. 14 is a schematic diagram of the centrifugal pump pumping device. The functions of each part are described as follows.

(1) The filter screen and bottom valve are installed in the inlet of the suction pipe. The filter is used to prevent impurities from being absorbed into the pump. The bottom valve is a one-way valve to prevent leakage during filling before starting the pump. The bottom valve has a large head loss. If the vacuum pump is used to pump air and fill water, the bottom valve can be cancelled.

(2) Layout of inlet pipes. Inlet pipes should avoid air storage in the pipes and strictly prevent air intake during installation.

(3) Eccentric diffuser tube gradually expanding pipes are used to enlarge the intake

pipes so as to reduce head loss in the pipes.

(4) The vacuum gauge is installed at the pump inlet to measure the vacuum value of the inlet.

(5) The pressure gauge is installed at the pump outlet to measure the outlet pressure.

(6) The check valve is installed in the outlet pipe near the water pump and is a one-way valve, which closes automatically when the accident stops, preventing the water from flowing back and avoiding the high-speed reversal of the unit. However, the installation of the check valve will not only increase the head loss, but also produce a large water hammer pressure due to its sudden closing. In order to cancel the check valve, a flap valve can be set at the outlet or other means can be used to cut off the flow. Slow-closing check valve can also be used instead of burst-closing check valve.

(7) After the gate valve is installed in the check valve, it can be used to regulate the flow and power of the pump, start the valve to reduce the starting load of the power engine.

(8) Flap valve is a one-way valve device to prevent water from flowing back into the pipe in the outlet sump when the door is tapped to stop operation.

Not all of the above-mentioned pumping equipment and accessories are required. A choice should be made according to the specific conditions. Among the three valves (gate valve, check valve and bottom valve), check valve and bottom valve have great head loss and energy consumption, so they should be cancelled.

2.3.2   Axial flow pump pumping device

The impeller of small and medium-sized vertical axial flow pump is installed under the water surface of the intake sump. When running, the power machine drives the impeller to rotate in the water. After the water flow enters the impeller from the bell mouth, it enters the outlet sump through the guide vane, the outlet elbow and the outlet pipe. Because the impeller is submerged, there is no need to fill the water before starting the pump, so there is no need to set a bottom valve. Axial flow pump is not allowed to close valve to start, and gate valve shall not be installed on the outlet pipe. To prevent water backflow during shutdown, flap valves or other shutoff devices are set at the outlet of the outlet pipe.

Fig. 2. 15 presents a small and medium-sized vertical axial flow pump pumping device.

2. 4   Performance parameters of water pump

The performance parameters of the pump include the flowrate (discharge), head, speed, power, efficiency, Net Positive Suction Head (suction vacuum height), etc. Among the six parameters, flowrate, head and speed are the basic parameters. As long as one of them changes, the other parameters will change more or less according to certain rules. However, under the drive of the power engine, the speed of the pump is generally fixed at a certain value, so the mutually changing law of the other five parameters can be used to reflect the performance of the pump at this speed.

2.4.1   Flowrate

Pump flowrate is divided into volume flowrate and mass flowrate. Volume flowrate is the volume of water pumped by the pump in a unit time, and its unit is m^3/s, L/s or $m^3/$ h. Mass flowrate is the mass of the water pumped by the pump in a unit time in kg/s or t/ h. Volume flowrate is usually used and is represented by the symbol Q.

2.4.2   Head

The head also known as water head, is the total energy transmitted by the impeller to the unit weight of water by the pump, which is the concept of energy. It can be expressed by the difference of the total energy E_1 and E_2 per unit on the section of the pump inlet and outlet, and its unit is measured water column height m.

Taking the horizontal centrifugal pump shown in Fig. 2.16 as an example, and taking the axis of the pump as the reference plane, then

$$E_1 = z_1 + \frac{p_1}{\rho g} + \frac{v_1^2}{2g}, \quad E_2 = z_2 + \frac{p_2}{\rho g} + \frac{v_2^2}{2g}$$

$$H = E_2 - E_1 = (z_2 - z_1) + \frac{p_2 - p_1}{\rho g} + \frac{v_2^2 - v_1^2}{2g} \tag{2.1}$$

Where, z_1 and z_2 are the vertical distance from vacuum gauge pressure measuring point and zero position of pressure gauge to datum surface respectively, and negative value (m) is taken when it is lower than datum surface; p_1, p_2 are absolute pressure head (m) of vacuum gauge pressure measuring point and zero position of pressure gauge respectively; $\frac{v_1^2}{2g}$, $\frac{v_2^2}{2g}$ are the flow head (m) of the inlet and outlet sections of the pump respectively but the reading of gauge is relative value, substitute the above formula as

$$H = E_2 - E_1 = (z_2 - z_1) + (H_s + H_d) + \frac{v_2^2 - v_1^2}{2g} \tag{2.2}$$

Where, H_s、H_d are vacuum gauge and pressure gauge readings (m) respectively.

For the head calculation of vertical axial flow pump, if the impeller is submerged in water, then $H_s = 0$. When the following sump surface is the reference level, $z_1 = 0$, v_1 is very small or negligible. See Fig. 2.17, then

$$H = z_2 + H_d + \frac{v_2^2}{2g} \tag{2.3}$$

2.4.3   Power

There are two types of pump power.

(1) The effective power N_e is the net power actually obtained by the water in the pump, which can be calculated according to the flowrate and head.

$$N_e = \frac{\rho g Q H}{1000} \tag{2.4}$$

(2) Shaft power N_a. The external power required by the pump in a certain flowrate and head, that is, the power transmitted by the power engine to the shaft of the pump, is also known as the input power (kW). It is impossible to transmit all the shaft power to the water, but the remaining part will become effective power after losing part of the power.

2. 4. 4　Efficiency

The ratio of effective power to shaft power is efficiency

$$\eta = \frac{N_e}{N_a} \times 100\% \tag{2.5}$$

The pump efficiency reflects the effective degree of the power delivered by the pump, which reflects the value of the power loss in the pump, and is an important technical and economic indicator to measure the performance of the pump. It is composed of three partial efficiencies in the pump: hydraulic efficiency, mechanical efficiency and volumetric efficiency.

(1) Mechanical loss N_{ml} and mechanical efficiency η_m. Mechanical loss includes friction loss between pump shaft and bearing, friction loss between pump shaft and packing in shaft seal, loss caused by impeller rotating in water, i. e. disc loss. After overcoming the mechanical loss of power N_{ml}, the pump transfers the remaining power to the pumped water. This part of the power is called water power N_h. The mechanical efficiency η_m is

$$\eta_m = \frac{N_a - N_{ml}}{N_a} = \frac{N_h}{N_a} \times 100\% \tag{2.6}$$
$$N_h = \rho g Q_T H_T$$

Where, Q_T is the theoretical flowrate of the pump (m^3/s); H_T is the theoretical head of the pump (m).

(2) Volume loss and volume efficiency η_V. Volume loss in the pump refers to power loss caused by leakage from the high pressure side to the low pressure side. In the water body pumped by the pump, the partial flowrate outside the pump is expressed by Q, and the leakage flowrate is expressed by q. Therefore, there is $Q_T = Q + q$. The energy consumed by the leakage is $N_q = \rho g q H_T$, so that $N' = N_h - N_q = \rho g Q H_T$, and the volume efficiency is

$$\eta_V = \frac{N'}{N_h} = \frac{\rho g Q H_T}{\rho g Q_T H_T} = \frac{Q}{Q_T} \times 100\% \tag{2.7}$$

(3) Hydraulic efficiency η_h. The hydraulic loss h caused by friction resistance, whirlpool and impact in the over-flow components such as suction chamber, vane groove, passage and outlet chamber of the pump. The hydraulic efficiency is as follows

$$\eta_h = \frac{N_e}{N'} = \frac{N' - N_h}{N'} = \frac{\rho g Q H}{\rho g Q H_T} = \frac{H}{H_T} \times 100\% \tag{2.8}$$

According to Eq. (2.5) – Eq. (2.8), then

$$\eta = \eta_m \eta_V \eta_h \tag{2.9}$$

The efficiency of a water pump is equal to the product of the mechanical efficiency, volume efficiency and hydraulic efficiency of the pump. To improve the efficiency of the pump, it is necessary to reduce the mechanical friction and leakage as much as possible, and strive to improve the hydraulic optimization design of flow passage parts and improve the quality of manufacturing and assembly.

2.4.5   Speed

Rotational speed n refers to the number of rotations per minute of the impeller. The rated speed indicated on the nameplate of the pump is the speed under design conditions. When the speed changes, the working performance of the pump will also change.

2.4.6   Suction vacuum height and *NPSH* (net positive suction head)

These two parameters are the parameters reflecting the water suction performance or cavitation performance of the pump. They are the main parameters to determine the installation height of the pump and to evaluate the cavitation problem of the pump (See Chapter 6 for details on pump cavitation performance).

第3章　泵内流动分析与基本理论

　　水体在泵内的流动十分复杂，对泵的性能影响很大，历来是水力机械的研究重点。本章从分析叶轮内流动出发，阐明水泵叶轮对水体做功的机理，从而推导出基本方程，为分析水泵应用中存在的技术问题以及水泵设计选型提供理论基础。

3.1　叶轮内流动分析

　　实际水体在叶轮中的流动属于三维复杂流动。为便于分析流态，将离心泵叶轮分解为平面投影和轴面投影（图 3.1）。这样，将复杂的三维流动简化为二维流动，并假定流动是在完全平滑的叶槽中进行，暂不考虑水的黏性和壁面粗糙度的影响。当叶轮旋转时，水体通过叶槽的同时也被叶轮强迫着一起旋转，因此在叶槽中的每个水质点，既对于叶轮做相对运动又随着叶轮做牵连运动。这两种运动的合成，就是水质点对于固定泵壳所做的绝对运动。图 3.2 所示为水体运动迹线在垂直轴线平面上的投影。

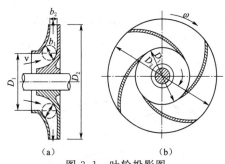

图 3.1　叶轮投影图

Fig. 3.1　Projection of impeller

（a）轴面投影；（b）平面投影

（a）Meridional view；（b）Planar view

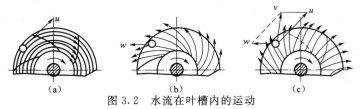

图 3.2　水流在叶槽内的运动

Fig. 3.2　Flow motion in blade groove

（a）牵连运动；（b）相对运动；（c）绝对运动

（a）Involved motion；（b）Relative motion；（c）Absolute motion

3.2　泵内流动理论模型

　　欧拉假设叶片式流体机械的流体为理想流体，流体模型为轴对称微元流束。欧拉证实了叶轮在等速旋转和稳定流的情况下，微元束所产生并传递到叶片上的力矩，等于动量矩的变化量。据此，欧拉推导出了著名的叶片泵基本方程。欧拉定理一直是叶片泵重要的理论基础。

由图 3.3 可知，叶轮中相对速度为 $\vec{w}$，牵连速度（圆周速度）为 $\vec{u}$，绝对速度为 $\vec{v}$，则

$$\vec{v} = \vec{u} + \vec{w} \tag{3.1}$$

图中绝对液流角 α 是绝对速度 $\vec{v}$ 与牵连速度 $\vec{u}$ 的夹角，相对液流角 β 是相对速度 $\vec{w}$ 与 $-\vec{u}$ 的夹角。绝对速度 $\vec{v}$ 又可分解为两个相互垂直的分速度：圆周分速 $\vec{v}_u$ 和轴面分速 $\vec{v}_m$。

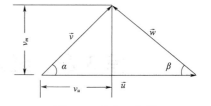

图 3.3　速度三角形

Fig. 3.3　Velocity triangle

$$\vec{v} = \vec{v}_u + \vec{v}_m \tag{3.2}$$

圆周分速 $\vec{v}_u$ 与圆周速度的方向一致，轴面分速 $\vec{v}_m$ 则是绝对速度 $\vec{v}$ 在过该质点的轴面内投影。在离心泵中，若不计轴向速度，$\vec{v}_m$ 就是绝对速度的径向分速度（图 3.4）；在轴流泵中，若不计径向速度，$\vec{v}_m$ 就是绝对速度的轴向分速度（图 3.5）。

速度三角形适用于叶槽流场中任何一点，叶轮进口速度三角形和出口速度三角形的参数分别用下标"1"和"2"来表示。通常假定泵的进口水流无旋，即 $v_{u1} = 0$，$\alpha_1 = 90°$；出口水流相对速度 w_2 方向与叶轮出口切线方向一致。

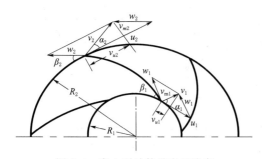

图 3.4　离心泵叶轮进出口速度

Fig. 3.4　Velocity of the inlet and outlet of the centrifugal pump impeller

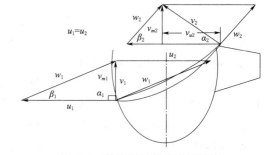

图 3.5　轴流泵叶轮进出口速度

Fig. 3.5　Velocity of the inlet and outlet of the axial flow pump impeller

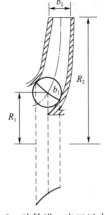

图 3.6　叶轮进、出口过水断面

Fig. 3.6　Flow section at inlet and outlet of impeller

若已知叶轮流道的几何形状、流量和转速，可按下述步骤作出进、出口速度三角形。

（1）进、出口圆周速度 u_1、u_2。已知叶轮进、出口半径为 R_1、R_2，转速为 n，则

$$u_1 = \frac{\pi R_1 n}{30}, \quad u_2 = \frac{\pi R_2 n}{30} \tag{3.3}$$

（2）进、出口绝对速度的轴面分速 v_{m1}、v_{m2}。

$$v_{m1} = \frac{Q_T}{A_1}, \quad v_{m2} = \frac{Q_T}{A_2} \tag{3.4}$$

式中：Q_T 为流经叶轮的理论流量；A_1、A_2 为叶轮进、出口过水断面面积（图 3.6）。

$$A_1 = 2\pi R_1 b_1 \Psi_1, \quad A_2 = 2\pi R_2 b_2 \Psi_2 \tag{3.5}$$

式中：b_1、b_2 为叶轮进、出口处流道宽度；Ψ_1、Ψ_2 为叶轮进、

出口处考虑叶片厚度时的排挤系数，离心泵计算时 $\Psi_1 = 0.75 \sim 0.88$，$\Psi_2 = 0.85 \sim 0.95$，小泵取小值，大泵取大值。

将式（3.5）代入式（3.4），得

$$v_{m1} = \frac{Q_T}{2\pi R_1 b_1 \Psi_1 A_1}, \quad v_{m2} = \frac{Q_T}{2\pi R_2 b_2 \Psi_2 A_2} \tag{3.6}$$

（3）进口绝对液流角 α_1。由叶轮的进口形式决定，水泵在设计状态下运行时，$\alpha_1 = 90°$。

（4）出口相对速度 w_2。w_2 的方向为叶片出口末端的切线方向。

有了以上各参数，则叶轮进、出口速度三角形就可以求出。

3.3 水泵的基本方程

水泵基本方程通过建立叶轮对水体做功与水体运动状态之间的关系推导得出水泵的理论扬程。

3.3.1 基本方程的推导

在建立叶轮泵的基本方程时，假定：①无黏性理想流体；②定常流动；③叶轮具有无限多、无限薄的叶片。

根据动量矩定理，液体质点系对于任一轴的动量矩 L 随时间 t 的变化率，等于作用于该质点系所有外力对同一轴的力矩之和 M，可用下式表达

$$\frac{\mathrm{d}L}{\mathrm{d}t} = M \tag{3.7}$$

把动量矩定理应用于离心泵叶轮的一个叶槽内的水流。如图 3.7 所示，在叶轮上取某一叶槽，在 $t = 0$ 时，槽内水流居于 $abcd$ 的位置，经过 $\mathrm{d}t$ 时间以后，这部分水体位置变为 $efgh$，这部分水体对泵轴的动量矩的变化量是两个位置动量矩之差，即

$$\mathrm{d}L = L_{efgh} - L_{abcd} \tag{3.8}$$

$\mathrm{d}t$ 时间内，尚在叶槽内的水体为 $abgh$，流入叶轮的水体为 $hgcd$，流出叶槽的水体为 $efba$，故该质点系的动量矩变化应等于这两部分水体动量矩之差，即

$$\mathrm{d}L = L_{efba} - L_{hgcd} \tag{3.9}$$

在恒定流状态下，$\mathrm{d}t$ 时段内流出叶槽的水体与流入叶槽的水体具有相等的质量 $\mathrm{d}m$，故两块水体对泵轴的动量矩分别为

$$L_{hgcd} = \mathrm{d}m\, v_1 \cos\alpha_1 R_1, \quad L_{efba} = \mathrm{d}m\, v_2 \cos\alpha_2 R_2 \tag{3.10}$$

$$\mathrm{d}L = \mathrm{d}m\,(v_2 \cos\alpha_2 R_2 - v_1 \cos\alpha_1 R_1) \tag{3.11}$$

因此由式（3.7），叶槽内的水流动量方程为

$$M_{pa} = \frac{\mathrm{d}m}{\mathrm{d}t}(v_2 \cos\alpha_2 R_2 - v_1 \cos\alpha_1 R_1) \tag{3.12}$$

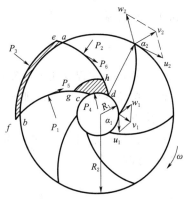

图 3.7 叶槽中液流瞬时变化状况及作用力

Fig. 3.7 Transient change of fluid flow and force in the impeller channel

M_{pa} 为作用在叶槽内水体上所有的外力矩，作用于整股水体上的外力如下（图 3.7）：

（1）叶片迎水面和背水面作用于水体的压力 P_1 及 P_2。

（2）作用于 ab 和 cd 面上的水压力 P_3 及 P_4，由于力的方向均为径向，所以对泵轴的力矩为零。

（3）作用于水流的摩阻力 P_5 和 P_6，理想流体计算时不考虑。

把式（3.12）推广应用到流过叶轮的全部水体时，式中的 M 变为作用于全部水体的所有外力矩之和，而在稳定流动的假设下有

$$\sum M_{pa} = \int \frac{\mathrm{d}m}{\mathrm{d}t}\ (v_2 \cos\alpha_2 R_2 - v_1 \cos\alpha_1 R_1)$$

其中

$$\int \frac{\mathrm{d}m}{\mathrm{d}t} = \int \frac{\mathrm{d}(\rho V)}{\mathrm{d}t} = \rho \int \frac{\mathrm{d}V}{\mathrm{d}t} = \rho \int \mathrm{d}Q_T = \rho Q_T$$

$$M = \rho Q_T\ (v_2 \cos\alpha_2 R_2 - v_1 \cos\alpha_1 R_1) \tag{3.13}$$

根据假设，不计黏性阻力损失，叶轮轴功率 N 全部传给水体，叶轮轴功率为

$$N = \rho g Q_T H_T \tag{3.14}$$

又

$$N = M\omega \tag{3.15}$$

式中：ω 为叶轮旋转角速度。

根据式（3.14）和式（3.15）可得

$$H_T = \frac{M\omega}{\rho g Q_T} \tag{3.16}$$

将式（3.13）代入式（3.16），整理后有

$$H_T = \frac{\omega}{g}\ (v_2 \cos\alpha_2 R_2 - v_1 \cos\alpha_1 R_1) \tag{3.17}$$

又

$$\omega r = u,\ v \cos\alpha = v_u$$

故

$$H_T = \frac{1}{g}\ (u_2 v_{u2} - u_1 v_{u1}) \tag{3.18}$$

式（3.18）是在无穷多叶片假设下推导出来的，所以将 H_T 下标加上"∞"，这样就得到叶片泵的基本方程式

$$H_{T\infty} = \frac{1}{g}\ (u_2 v_{u2} - u_1 v_{u1}) \tag{3.19}$$

水泵的基本方程反映了叶轮对液体所做的功与液体运动的关系，表明叶轮在动力机驱动下传给单位液体的能量，即产生的扬程，其大小与叶轮旋转速度和叶轮出口速度的圆周分量成比例。

式（3.19）是以无穷多叶片的假设为前提的，由于大多数情况下 $v_{u1} = 0$，故式（3.19）可改写为

$$H_{T\infty} = \frac{1}{g} u_2 v_{u2} \tag{3.20}$$

当叶片为有限多时，则引入修正系数 K，得

$$H_T = K H_{T\infty} \tag{3.21}$$

此修正公式仅适用叶片数较多的离心泵，而轴流泵由于叶片数较少、通道很宽等原

因，以至叶片有限多的修正公式不再适用，由此需要升力理论来解决。

3.3.2 基本方程的讨论

（1）基本方程式只与叶轮进、出口的动量矩有关，与叶片的形状无关。不论叶轮内部水流方式如何，能量的传递都决定于进、出口速度三角形，基本方程式既适用于离心泵，也适用于轴流泵、混流泵等叶片泵。

（2）基本方程式与被抽送的液体种类无关，适合于一切液体和气体，只是 H_T 应当用被抽液体的米液（气）柱高度计。

（3）水泵扬程主要取决于出口速度图，因为大多数情况下 $v_{u1}=0$。

（4）轴流泵叶片扭曲形状与速度三角形有关。轴流泵转动时，叶轮距泵轴不等的断面有不同的速度三角形。如果用半径 $r_内$ 及 $r_外$ 的两个圆柱面横切叶片，便得到内外两个叶片剖面，这两个剖面在设计工况下的出口速度三角形如图 3.8 所示。当叶轮以角速度 ω 旋转时，$u_{2内}=r_1\omega$，$u_{2外}=r_2\omega$，由于 $r_2>r_1$，所以 $u_{2外}>u_{2内}$。设计要求叶片内外两个断面所产生的扬程必须相等，即

$$\frac{u_{2外}v_{u2外}}{g}=\frac{u_{2内}v_{u2内}}{g} \tag{3.22}$$

要满足上式必须使 $v_{u2外}<v_{u2内}$，从图 3.8 可以看出，只有当 $\beta_外<\beta_内$ 才能满足设计要求。因此离泵轴越远的剖面，安放角越小，这就是轴流泵叶片扭曲形状的原因。

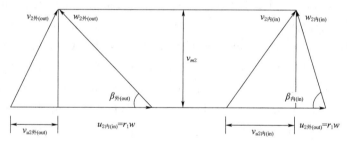

图 3.8　内外断面出口速度三角形

Fig. 3.8　Outlet velocity triangle of inner and outer sections

Chapter 3   Flow Principle and Fundamental Theory of Pump

The flow of water in the pump is very complex, which has a great impact on the performance of the pump, and has always been the focus of the study of hydraulic machinery. In this chapter, starting from the analysis of the flow inside the impeller, the mechanism of the impeller working on the water body is elucidated, and the basic equation is deduced, which provides a theoretical basis for the analysis of the technical problems existing in the application, the design and selection of the pump.

3.1   Flow inside impeller

The actual flow in impeller belongs to three-dimensional complex flow. To facilitate the analysis of flow regime, centrifugal pump impeller is decomposed into plane projection and meridional projection for study (see Fig. 3.1). In this way, the complex three-dimensional flow is simplified to two-dimensional flow, and it is assumed that the flow is carried out in a completely smooth blade groove, without considering the influence of water viscosity and wall roughness for the moment. When the impeller rotates, the water body is forced to rotate together by the impeller while passing through the blade groove. Therefore, at each water particle in the blade groove, the relative motion of the impeller and the implicated motion of the impeller are made. The combination of these two movements is the absolute motion of the water particle for the fixed pump housing. Fig. 3.2 shows the projection of water motion trace on the plane of vertical axis.

3.2   Theoretical model of flow in pump

Euler assumes that the fluid in the vane fluid machine is an ideal fluid and the fluid model is an axisymmetric micro-element flow beam. Euler confirms that the moment generated by the bundle of micro-elements and transmitted to the blade is equal to the change of momentum moment when the impeller rotates at constant speed and steady flow. Accordingly, Euler deduced the well-known basic equation of vane pump. Euler Theorem has always been an important theoretical basis for vane pumps.

Fig. 3.3 shows the velocity triangles of blade. Let the relative velocity be $\vec{w}$, the implicated velocity (circumferential velocity) be $\vec{u}$, and the absolute velocity be $\vec{v}$, then

$$\vec{v} = \vec{u} + \vec{w} \tag{3.1}$$

In the Fig. 3. 3, the absolute liquid flow angle α is the angle between the absolute velocity $\vec{v}$ and the implicated velocity $\vec{u}$, and the relative liquid flow angle β is the angle between the relative velocity $\vec{w}$ and $-\vec{u}$. The absolute velocity $\vec{v}$ can be decomposed into two mutually perpendicular velocities: the circumferential velocities $\vec{v}_u$ and the axial velocities $\vec{v}_m$.

$$\vec{v} = \vec{v}_u + \vec{v}_m \tag{3.2}$$

The tangential component of absolute velocity $\vec{v}_u$ is consistent with the direction of the circumferential velocity, and the axial velocity $\vec{v}_m$ is the projection of the absolute velocity in the axial plane over the particle. In centrifugal pumps, $\vec{v}_m$ is the radial partial velocity of absolute velocity regardless of the axial velocity (see Fig. 3. 4). In axial flow pumps, $\vec{v}_m$ is the axial partial velocity of absolute velocity regardless of the radial velocity (see Fig. 3. 5).

The velocity triangle is applicable to any particle in the flow field of the blade channel, and the parameters of the impeller inlet velocity triangle and outlet velocity triangle are expressed by the subscripts "1" and "2", respectively. It is generally assumed that the inlet flow of the pump is irrotational, i. e. $v_{u1} = 0$, $\alpha_1 = 90°$; and the direction of relative velocity w_2 of the outlet flow is consistent with the tangential direction of the impeller outlet.

If the geometry, flowrate and speed of the impeller passage are known, the inlet and outlet velocity triangles can be made according to the following steps.

(1) Circumferential velocity of inlet and outlet u_1 and u_2. If the radii of inlet and outlet of impellers are known to be R_1, R_2 and speed n, then

$$u_1 = \frac{\pi R_1 n}{30}, \quad u_2 = \frac{\pi R_2 n}{30} \tag{3.3}$$

(2) Axial velocity of absolute inlet and outlet velocity v_{m1}、v_{m2}.

$$v_{m1} = \frac{Q_T}{A_1}, \quad v_{m2} = \frac{Q_T}{A_2} \tag{3.4}$$

Where, Q_T is the theoretical flowrate through the impeller; A_1 and A_2 are the cross-section areas of inlet and outlet water flow through the impeller (see Fig. 3. 6).

$$A_1 = 2\pi R_1 b_1 \Psi_1, \quad A_2 = 2\pi R_2 b_2 \Psi_2 \tag{3.5}$$

Where, b_1 and b_2 are the width of the flow passage at the inlet and outlet of the impeller; Ψ_1 and Ψ_2 are the extrusion coefficients when considering the blade thickness at the inlet and outlet of impeller. $\Psi_1 = 0.75 - 0.88$, $\Psi_2 = 0.85 - 0.95$ when calculating centrifugal pump, small pump takes small value and large pump takes large value.

Put Eq. (3.5) into Eq. (3.4) to obtain

$$v_{m1} = \frac{Q_T}{2\pi R_1 b_1 \Psi_1 A_1}, \quad v_{m2} = \frac{Q_T}{2\pi R_2 b_2 \Psi_2 A_2} \tag{3.6}$$

(3) Absolute inlet liquid flow angle α_1. α_1 is determined by the inlet form of the impeller. When the pump runs at a design state, α_1 equal ninety degree.

(4) Relative velocity at outlet w_2. The direction of w_2 is tangential to the end of blade outlet.

With the above parameters, the inlet and outlet velocity triangles of the impeller can be obtained.

3.3   Basic equation of impeller pump

The basic equation of the pump deduces the theoretical head of the pump by establishing the relationship between the work of impeller on liquid and the state of liquid motion.

3.3.1   Derivation of the basic equation of impeller pumps

When establishing the basic equation of impeller pumps, the following assumptions are made: ①inviscid ideal fluid; ②steady flow; ③the number of impeller blades is infinite, or the blades are infinitely thin.

According to the theorem of momentum moment, the rate of change of momentum moment L of a liquid particle system with time t for any axis is equal to the sum of the moments M of all external forces acting on the same axis of the particle system. It can be expressed as follows

$$\frac{\mathrm{d}L}{\mathrm{d}t} = M \tag{3.7}$$

The theorem of momentum moment is applied to the flow in a impeller of the centrifugal pump impeller. As shown in Fig. 3.7, when a blade channel is taken on the impeller, at $t=0$, the water in the channel resides in the position of $abcd$. After $\mathrm{d}t$ time, the position of this part of water becomes $efgh$, and the change of momentum moment of this part of water to the pump shaft is the difference of momentum moment of two positions

$$\mathrm{d}L = L_{efgh} - L_{abcd} \tag{3.8}$$

In $\mathrm{d}t$ time, the water still in the blade channel is $abgh$, the water flowing into the impeller is $hgcd$, and the water flowing out of the blade channel is $efba$. Therefore, the change of momentum moment of this particle system should be equal to the difference of momentum moment between the two parts of water

$$\mathrm{d}L = L_{efba} - L_{hgcd} \tag{3.9}$$

Under constant flow condition, the water flowing out of the channel and the water flowing into the channel have the same mass $\mathrm{d}m$ during $\mathrm{d}t$ period, so the momentum moments of the two water bodies to the pump shaft are respectively

$$L_{hgcd} = \mathrm{d}m v_1 \cos\alpha_1 R_1, \quad L_{efba} = \mathrm{d}m v_2 \cos\alpha_2 R_2 \tag{3.10}$$

$$\mathrm{d}L = \mathrm{d}m \ (v_2 \cos\alpha_2 R_2 - v_1 \cos\alpha_1 R_1) \tag{3.11}$$

Therefore, from Eq. (3.7), the equation of water in the blade channel is

$$M_{pa} = \frac{dm}{dt} \ (v_2 \cos\alpha_2 R_2 - v_1 \cos\alpha_1 R_1) \tag{3.12}$$

M_{pa} is all the external moments acting on the water in the channel, and the external forces acting on the whole water area are as follows (see Fig. 3.7):

(1) Pressures P_1 and P_2 acting on the water area at the blade upstream and backwater surfaces.

(2) The water pressures P_3 and P_4 acting on the ab and cd surfaces all act radially, so the moment to the pump shaft is zero.

(3) The friction resistance P_5 and P_6 acting on water area is not considered in the derivation.

When Eq. (3.12) is extended to whole water bodies flowing through the impeller, M in equation becomes the sum of all external moments acting on whole water area, and under the assumption of steady flow there are

$$\sum M_{pa} = \int \frac{dm}{dt} \ (v_2 \cos\alpha_2 R_2 - v_1 \cos\alpha_1 R_1)$$

Which
$$\int \frac{dm}{dt} = \int \frac{d(\rho V)}{dt} = \rho \int \frac{dV}{dt} = \rho \int dQ_T = \rho Q_T$$

$$M = \rho Q_T \ (v_2 \cos\alpha_2 R_2 - v_1 \cos\alpha_1 R_1) \tag{3.13}$$

According to the assumption, regardless of the resistance loss, the impeller shaft power N is all transmitted to the water, and the shaft power of impeller is

$$N = \rho q Q_T H_T \tag{3.14}$$

and
$$N = M\omega \tag{3.15}$$

Where, ω is the rotation angle speed of the impeller.

According to the Eq. (3.14) and Eq. (3.15)

$$H_T = \frac{M\omega}{\rho g Q_T} \tag{3.16}$$

Substituting Eq. (3.13) into Eq. (3.16), the results are as follows

$$H_T = \frac{\omega}{g} \ (v_2 \cos\alpha_2 R_2 - v_1 \cos\alpha_1 R_1) \tag{3.17}$$

and
$$\omega r = u, \quad v\cos\alpha = v_u$$

so
$$H_T = \frac{1}{g} \ (u_2 v_{u2} - u_1 v_{u1}) \tag{3.18}$$

The above equation is derived under the assumption of infinite number of vanes, so the H_T subscript is added with "∞", which leads to the basic equation of the impeller pump

$$H_{T\infty} = \frac{1}{g} \ (u_2 v_{u2} - u_1 v_{u1}) \tag{3.19}$$

The basic equation of the pump reflects the relationship between the work done by the impeller to the liquid and the movement of the liquid, it shows that the energy transmitted by the impeller to the unit liquid under the driving of the power machine, promptly the generated head, whose value is proportional to the circumferential component of the impeller rotating speed and the impeller outlet velocity.

Eq. (3. 19) is based on the assumption of infinite number of blades. Since in most cases, $v_{u1} = 0$, Eq. (3. 19) can be rewritten as

$$H_{T\infty} = \frac{1}{g} u_2 v_{u2} \tag{3.20}$$

When the number of blades is limited, a correction factor should be introduced to obtain

$$H_T = K H_{T\infty} \tag{3.21}$$

The above correction formula is only applicable to centrifugal pumps with more blades, while the axial flow pumps are not applicable because of the small number of blades and the wide passage, so that the correction formula with limited blades is no longer applicable, which needs lifting theory to solve.

3. 3. 2 Discussion of basic equations

(1) The basic equation is only related to the momentum moment at the inlet and outlet of the impeller, but not to the shape of the blade. Regardless of the flow pattern inside the impeller, the energy transfer is determined by the inlet and outlet velocity triangles. The basic equation is applicable to both centrifugal pumps and vane pumps such as axial and mixed flow pumps.

(2) The basic equation is suitable for all liquids and gases, irrespective of the type of liquid being pumped, except that H_T should be measured with a metre liquid (gas) column altimeter of the liquid being pumped.

(3) The pump head mainly depends on the exit velocity chart, because in most cases $v_{u1} = 0$.

(4) The axial flow pump blade distortion shape is related to velocity triangle. When the axial flow pump is rotating, there are different velocity triangles in the cross-section of the impeller with different distance from the pump shaft. If the two cylindrical surfaces r_{in} and r_{out} are used to cross cut the blade, the internal and external blade sections will be obtained. The outlet velocity triangles of these two sections under the design conditions are shown in Fig. 3. 8.

When the impeller rotates at angular speed ω, u_{2in} equals $r_1\omega$, u_{2out} equals $r_2\omega$, because r_2 is larger than r_1, so u_{2out} is larger than u_{2in}. The design requires that the head generated by the inner and outer sections of the blade be equal, i. e.

$$\frac{u_{2out} v_{u2out}}{g} = \frac{u_{2in} v_{u2in}}{g} \tag{3.22}$$

To satisfy the above formula, it is necessary to make $v_{u2out} < v_{u2in}$. it can be seen from Fig. 3. 8, that the design requirements can only be met when $\beta_{out} < \beta_{in}$. Therefore, the reason for the twisted shape of axial flow pump blades is that the farther the section away from the pump shaft, the smaller the placement angle.

第 4 章 水泵性能及相似理论

水泵的 6 个工作参数代表了水泵的性能，但各个工作参数不是孤立的，而是有一定的内在联系和变化规律。如将它们的变化规律用一组曲线表示，这组曲线就称为性能曲线。通常在水泵转速为某一定值时，根据扬程、轴功率、效率、吸上真空高度或汽蚀余量随着流量而变化的关系绘制出各种性能曲线。

只有掌握水泵的性能，熟悉各种水泵性能曲线的特点，才能合理进行泵站水泵选型配套，正确确定水泵的安装高程以及解决水泵装置在运行中所遇到的许多问题。

4.1 理 论 性 能 曲 线

由于水体在叶片为有限多叶轮内的流动状况十分复杂，各种损失很难准确计算，目前还不能用理论方法确定性能曲线，只能从理论上做一些定性分析。

4.1.1 扬程与流量关系曲线

由水泵基本方程可知，水泵流量与扬程是密切相关的。在速度三角形中

$$v_{u2} = u_2 - v_{m2}\cot\beta_2, \quad v_{m2} = \frac{Q_T}{\pi D_2 b_2 \Psi_2}$$

代入基本方程可得

$$H_{T\infty} = \frac{u_2^2}{g} - \frac{u_2\cot\beta_2}{g\pi D_2 b_2 \Psi_2}Q_T \tag{4.1}$$

对于一台特定的水泵，π、D_2、b_2、Ψ_2、β_2 均为常数，当转速一定时，u_2 也为定值，式（4.1）表示扬程随流量变化是一个直线方程。在离心泵中，当叶片的 $\beta_2 < 90°$ 时，理论扬程 $H_{T\infty}$ 随 Q_T 的增大而减小。该直线在纵坐标为 H 的轴上交于 $\dfrac{u_2^2}{g}$，直线的斜率取决于 β_2 的值，如图 4.1 所示。

（1）当有限多叶片时，理论扬程采用 $H_T = K H_{T\infty}$ 进行修正，因此 $H_{T\infty}$ 与 Q_T 的关系仍然是一条直线，其在纵坐标 H 轴上的交点为 $\dfrac{u_2^2}{g}/K$。

（2）泵的实际扬程等于泵的理论扬程减去泵内的水力损失 h，h 包括摩擦损失 h_f 和局部损失 h_j。泵的实际扬程曲线为曲线 Ⅱ 的扬程减去 h，即曲线 Ⅲ（H-Q_T）。

（3）泵的容积损失与泵内压力有关，压力越大泄漏越大，其关系可用 H_T-q 曲线表示，在 H_T 较小（低压）时，泵内无泄漏，甚至水流还可以通过间歇过水，可能使流量表现为正值。由于泄漏实际流量减少，$Q = Q_T - q$，故须在曲线 H-Q_T 上相应的流量减去对应的 H_T-q 值，才得到所要求的 H-Q 曲线，即水泵理论扬程-流量关系性能曲线 Ⅳ

$(H - Q)$。

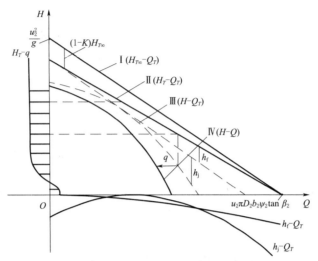

图 4.1　H - Q 曲线的推求

Fig. 4.1　Derivation of H - Q curves

4.1.2　功率-流量曲线

轴功率是水功率与机械损失功率之和。

水泵的理论输入功率 N_T 为

$$N_T = K\rho Q_T u_2 \left(u_2 - \frac{\cot\beta_2}{\pi D_2 b_2 \Psi_2} Q_T \right) \tag{4.2}$$

当 $\beta_2 < 90°$，图 4.2 上的曲线 I（N_T - Q_T）是一条向下弯曲的抛物线。

机械损失功率几乎与流量无关，曲线 I 上移 N_m 距离得到曲线 II（N - Q_T）。再将曲线 II 向左移动距离 q，得到最后的输入功率-流量曲线 III（N - Q）。

4.1.3　流量-效率曲线

从已知的 H - Q 曲线和 N - Q 曲线，可求得各对应流量下的效率值 η，即

$$\eta = \frac{\rho g Q H}{N_a}, \quad N_a > 0$$

当 $Q = 0$ 时，$\eta = 0$；当 $H = 0$ 时，$\eta = 0$，故水泵流量-效率曲线为一条通过坐标原点和横坐标上某一点的抛物线（图 4.2）。

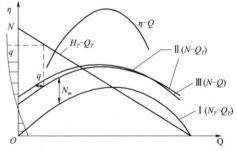

图 4.2　N - Q、η - Q 曲线的推求

Fig. 4.2　Derivation of N - Q and η - Q curves

4.2　水 泵 实 际 性 能

理论性能曲线只能进行定性分析，目前还难以定量确定，在实际中还必须用实测的方法来绘制性能曲线。水泵样本中绘制的性能曲线、性能表、型谱等，均为实测或根据模型

试验相似换算而来。

4.2.1 基本性能曲线

在水泵额定转速时或叶片设计安放角下，以 Q 为横坐标，以 H、N、η、H_s 或 $NPSH$ 为纵坐标，绘制的 H-Q、N-Q、η-Q、H_s-Q 或 $NPSH$-Q 曲线称为基本性能曲线。

图 4.3～图 4.5 所示为三种水泵的基本性能曲线。

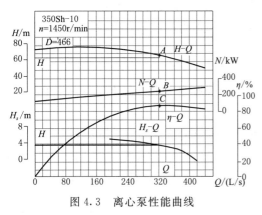

图 4.3　离心泵性能曲线

Fig. 4.3　Performance curve of centrifugal pump

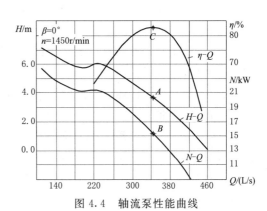

图 4.4　轴流泵性能曲线

Fig. 4.4　Performance curve of axial flow pump

从以上三种类型叶片泵的基本性能曲线，就可以看出它们的特点和差异。

1. 扬程-流量曲线（H-Q）

三种水泵的扬程-流量曲线，虽然都是下降的曲线，但离心泵下降平缓，轴流泵则是陡降。大多数轴流泵，在流量为设计工况的 40%～60% 时，出现马鞍形形状，即水力不稳定区。混流泵的扬程-流量曲线下降也较平缓，介于离心泵与轴流泵之间。

2. 功率-流量曲线（N-Q）

离心泵的功率-流量曲线具有上升的特点，如图 4.3 所示。零流量时的功率比设计工况的功率小得多。

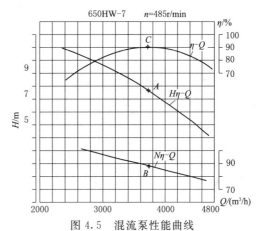

图 4.5　混流泵性能曲线

Fig. 4.5　Performance curve of mixed flow pump

轴流泵的功率-流量曲线与离心泵相反，具有下降的特点，在零流量时，功率达到最大值，可达设计工况功率的 2 倍左右。

混流泵的功率-流量曲线平坦，当流量变化时，功率变化很小，即使在零流量点，功率增加也较小，因此混流泵运行时平稳，具有较好的特性。

从功率-流量曲线的特点可知，离心泵应闭阀启动，使得动力机在小功率时平稳启动，待启动后再逐步打开闸阀。轴流泵则应开阀启动，一般轴流泵出水管上严禁安装闸阀，以免闭阀启动后功率严重超载，导致动力机损坏。混流泵闭阀或开阀启动根据启动方式而定。

3. 效率-流量曲线（η-Q）

三种水泵的效率-流量曲线总的趋势都是从最高效率点（BEP）向两边下降，但仍各有特点。离心泵、混流泵效率-流量曲线在最高效率点两侧变化平缓，高效区范围较宽，反映了离心泵、混流泵在流量变化较大的范围内，其效率变化不大，有利于流量调节。轴流泵的效率曲线在最高点两侧下降较陡、高效区范围较窄。

4. 允许吸上真空高度或临界汽蚀余量流量曲线（H_s-Q、$NPSH$-Q）

这两条曲线都是表示水泵吸水性能与汽蚀情况，但两者的变化规律截然不同，前者是一条下降的曲线，后者是具有最小极值的曲线。

4.2.2 通用性能曲线

把同一台泵不同转速或不同叶片角度时的性能曲线按照一定方式绘制在一张图上，就是水泵通用性能曲线。

图 4.6 所示是某台离心泵在转速为 900r/min、1000r/min、1100r/min 和 1200r/min 时的通用性能曲线。图 4.7 所示是叶片安装角为 $-6°$、$-4°$、$-2°$、$0°$、$+2°$、$+4°$时的轴流泵通用性能曲线。图中向下倾斜的实线曲线为扬程-流量关系曲线，等值线为等功率线和等效率线。

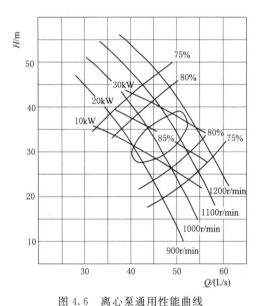

图 4.6　离心泵通用性能曲线

Fig. 4.6　General performance curve of centrifugal pump

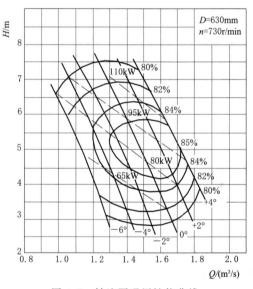

图 4.7　轴流泵通用性能曲线

Fig. 4.7　General performance curve of axial flow pump

通过水泵的通用性能曲线，可以方便地查出该泵在不同转速（或不同安装角）、不同流量时对应的其他参数，例如扬程、功率、效率和允许吸上真空高度（或允许汽蚀余量）。

4.2.3 性能表

在实际应用中，为方便快捷地查找水泵高效区的性能，以便选型，常用表格的形式表示水泵在某一转速下的性能参数，即性能表（表4.1）。

表 4.1 650 HW - 7 型混流泵性能表

Table 4.1   Performance table of 650 HW - 7 mixed flow pump

型号 Model of pump	流量 Discharge Q		扬程 Head H/m	转速 Speed n/(r/min)	功率 Power N/kW		效率 Efficiency η/%	临界汽蚀余量 $NPSH$/m
	m³/h	L/s			轴功率 Shaft power	配用功率 Matching power		
650 HW - 7	3060	850	7.4	450	72.5	100	85.0	5.3
	3400	944	6.5		68.0		88.0	
	3960	1100	5.0		63.4		85.0	
	3295	915	8.6	485	90.8	180	85.0	5.5
	3663	1017	7.6		86.1		88.0	
	4244	1185	5.9		80.6		85.0	

4.2.4 综合型谱图

将各种水泵通用性能曲线高效区范围内扬程和高效区效率围成的区域均绘制到同一张图上，就形成所谓的水泵综合型谱图，如图4.8所示。利用它可以快速方便地查找泵型，应用现代的信息技术，已建立有水泵泵型数据库，查选更为便利快捷。

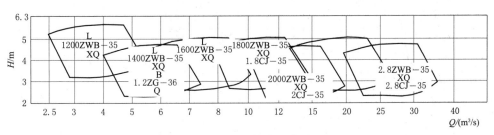

图 4.8 水泵综合型谱图

Fig. 4.8   Comprehensive type spectra of pump

4.2.5 水泵装置性能

前述的性能曲线、性能表、综合性能曲线、型谱等均为水泵性能，通常由水泵样本得到，但在实际泵站工程中，需要获得水泵装置的性能。水泵装置的扬程为装置进出口的能量差，与水泵扬程相比少进出水管路水头损失。图4.9所示为轴流泵抽水装置性能曲线。

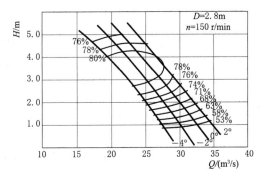

图 4.9 轴流泵抽水装置性能曲线

Fig. 4.9   Performance curves of axial flow pump system

4.2.6 全工况（四象限）性能

水泵正常运转时，其扬程、流量、转速和功率都是正值，但在反常情况下，部分或全部工作参数可能出现负值。

将性能曲线分为四个象限，前述的基本性能曲线为第一象限中的性能曲线。如果水泵的扬程、流量出现负值情况，则水泵的性能曲线就会超出第一象限的范围，而运行到第二、三、四象限中去，得到能够包括所有正常与反常情况下的水泵全工况性能曲线。全工况性能曲线同第一象限内水泵性能曲线一样，只能用试验的方法才能得到。

图 4.10 所示为试验获得的双吸泵的全工况性能曲线。全工况性能曲线在解决水泵水锤和飞逸转速问题上很有用。

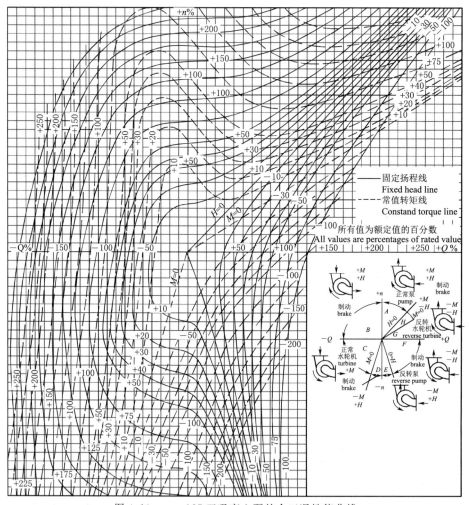

图 4.10　$n_s = 127$ 双吸离心泵的全工况性能曲线

Fig. 4.10　Complete characteristics of double-suction centrifugal pump，$n_s = 127$

4.3　水泵相似理论及应用

水泵的工作介质是液体，它在叶轮及流道中的运动受到各种力和边界条件的作用，呈现复杂的流动状态。单纯采用理论方法求解运动规律和工作性能存在一定的困难，需要用实验方法解决问题,即使用理论设计出新的水泵也需要采用模型试验进行验证。因此,实验方法是当今水力机械研究中广泛采用的一种手段。实验的理论基础是水力学相似理论。将相似理论应用到水泵的研究中来,可以解决以下问题:

(1) 根据模型试验成果,进行新泵的设计和制造。

(2) 对两台几何相似的水泵进行性能换算。

(3) 对同一台水泵进行不同转速下的性能换算。

因此,相似理论不仅用于水泵的设计和制造方面,还用于解决水泵运行中的问题。

4.3.1 相似条件

如果两台水泵保持水流流态相似,必须满足以下三个相似条件。

(1) 几何相似。即两台水泵的流道、叶轮形状和尺寸要相似,其糙率也要相似。在工艺上要保证糙率绝对值相等比较容易,但要保证相对糙率(Δ/D)相等比较困难。因此,在几何相似中,外形尺寸相似是主要的,其次才考虑糙率相似。

(2) 运动相似。即水流速度场相似,两台水泵中的水流相应质点上每对同名速度大小的比值为一常数,速度之间的夹角相等。

(3) 动力相似。即两台水泵内水流各相应质点所受的力,其性质和方向、大小成比例。作用在水泵内部水流的力包括重力、压力、惯性力和黏性力等。其中重力相似准则——弗劳德数相等;压力相似准则——欧拉数相等;惯性力相似准则——斯特劳哈尔数相等;黏滞力相似准则——雷诺数相等。

4.3.2 水泵相似律

水泵相似律就是两台泵在满足几何相似、运动相似和动力相似的前提下,它们的流量、扬程和轴功率之间遵循一定规律的变化关系。

1. 第一相似律

流过水泵叶轮的流量为 $Q=\pi Db\Psi v_m\eta_V$

则原型(P)、模型(M)流量比为

$$\frac{Q_P}{Q_M}=\frac{\pi D_P b_P \Psi_P v_{mP}\eta_{VP}}{\pi D_M b_M \Psi_M v_{mM}\eta_{VM}}$$

又

$$\frac{v_{mP}}{v_{mM}}=\frac{n_P D_P}{n_M D_M},\ \frac{b_P}{b_M}=\frac{D_P}{D_M}$$

当两泵的几何尺寸及转速相差不大时,设 $\Psi_P=\Psi_M$、$\eta_{VP}=\eta_{VM}$,则

$$\frac{Q_P}{Q_M}=\left(\frac{D_P}{D_M}\right)^3\frac{n_P}{n_M} \tag{4.3}$$

2. 第二相似律

泵的扬程为 $H=\dfrac{u_2 v_{u2}}{g}\eta_h$,则原型、模型扬程比为

$$\frac{H_P}{H_M}=\frac{u_{2P}v_{u2P}\eta_{hP}}{u_{2M}v_{u2M}\eta_{hM}}$$

因为

$$\frac{u_{2P}}{u_{2M}}=\frac{v_{u2P}}{v_{u2M}}=\frac{n_{1P}D_{1P}}{n_{1M}D_{1M}}$$

设 $\eta_{hP} = \eta_{hM}$，则

$$\frac{H_P}{H_M} = \left(\frac{D_P}{D_M}\right)^2 \left(\frac{n_P}{n_M}\right)^2 \tag{4.4}$$

3. 第三相似律

泵的轴功率为 $N = \dfrac{\rho g Q H}{\eta}$，则原型、模型功率比为

$$\frac{N_P}{N_M} = \frac{\rho g Q_P H_P \eta_M}{\rho g Q_M H_M \eta_P} = \frac{Q_P H_P}{Q_M H_M} \frac{\eta_M}{\eta_P}$$

将式（4.3）、式（4.4）代入上式，并设 $\eta_P = \eta_M$，则

$$\frac{N_P}{N_M} = \left(\frac{D_P}{D_M}\right)^5 \left(\frac{n_P}{n_M}\right)^3 \tag{4.5}$$

水泵相似律表明，若两台水泵相似，它们的流量之比与两泵线性尺寸比的三次方、转速比的一次方成正比；扬程比与线性尺寸比的平方、转速比的平方成正比；而轴功率之比，则与线性尺寸比的五次方、转速比的三次方成正比。

上述推导是以效率不变的假定为前提的，只有在转速和线性尺寸变化不太大的情况下，这一假定才能成立，故相似泵的大小和转速均有一定限制。

4.3.3　水泵比例律

对于同一台泵，因为 $\dfrac{D_P}{D_M} = 1$，根据水泵相似律，可得出

$$\frac{Q_1}{Q_2} = \frac{n_1}{n_2} \tag{4.6}$$

$$\frac{H_1}{H_2} = \left(\frac{n_1}{n_2}\right)^2 \tag{4.7}$$

$$\frac{N_1}{N_2} = \left(\frac{n_1}{n_2}\right)^3 \tag{4.8}$$

式中：下角标"1""2"分别表示水泵在不同转速下的工况。

比例律公式说明：当水泵的转速改变时，该泵的流量与转速成正比，扬程与转速的平方成正比，轴功率则与转速的三次方成正比。

4.3.4　水泵的性能换算

根据水泵的相似律或比例律，可以很方便地进行不同尺寸、转速的水泵性能换算。

必须注意的是，在推导水泵相似律时，假定相似水泵的效率相等，而实际上这两者是有差别的。另外，因无法做到完全的力学相似而带来的"比尺影响"，例如间隙、表面粗糙度等方面的影响，故相似换算的结果与实际情况有出入。对于要求较高的相似水泵性能换算，必须考虑上述影响，我国泵站模型试验验收规范有相关规定，可参阅有关文献。

4.4　比　转　速

由惯性力相似准则及压力相似准则，可得一个相似准则数

$$n_s = \frac{1000}{60} \frac{n\sqrt{Q}}{(gH)^{3/4}} \qquad (4.9)$$

由于习惯原因，将式（4.9）中的 g 去掉，并将该式乘以 3.65，变更成一个有因次的式（4.10），称为比转速。

$$n_s = \frac{3.65n\sqrt{Q}}{H^{3/4}} \qquad (4.10)$$

当模型泵的扬程为 1m，有效功率为 1hp（1hp＝0.7355kW），其流量可由下式求出

$$Q_s = \frac{1000N_a}{\rho g H_s} = \frac{1000\times0.7355}{1000\times9.81\times1} = 0.075 \ (\text{m}^3/\text{s}) \qquad (4.11)$$

这一模型泵的转速为 n_s，代入第三相似律，可求得式（4.10）。

在模型泵的转速前加一"比"字，是取比较之意。如把许多不同类型、不同形状的水泵都转化为各自的模型泵，且相关性能参数为扬程为 1m，有效功率为 1hp，流量为 0.075m³/s，则其水泵比转速 n_s 不可能相等。因此可以用它来对水泵进行比较和分类。根据大量实践经验可得到水泵泵型、几何形状及工作性能与比转速的关系，见表 4.2。

表 4.2　　　　　　　　　　水泵比转速与特性的关系
Table 4.2　　　Relationship between specific speed and characteristics of Pumps

水泵类型 pump type	离心泵 centrifugal pump			混流泵 mixed flow pump	轴流泵 axial flow pump
	低比转速 low specific speed	中比转速 middle specific speed	高比转速 high specific speed		
比转速 specific speed	50～80	80～150	150～300	300～500	500～1000
叶轮剖视简图 impeller section diagram					
尺寸比 size ratio	$\frac{D_2}{D_0}\approx2.5$	$\frac{D_2}{D_0}\approx2.0$	$\frac{D_2}{D_0}\approx1.8\sim1.4$	$\frac{D_2}{D_0}\approx1.2\sim1.1$	$\frac{D_2}{D_0}\approx0.8$
叶片形状 blade shape	圆柱形 cylindrical	进口扭曲形出口圆柱形 cylindrical outlet	扭曲形 twisted	扭曲形 twisted	扭曲形 twisted
工作性能曲线 performance characteristics					

由表 4.2 可见，随比转速从小到大的变化，水泵类型由离心泵—混流泵—轴流泵发生有规律的变化，离心泵的比转速为 50～300，混流泵的比转速为 300～500。近年混流泵比转速有增大的趋势，已达到 600 左右；轴流泵的比转速则大于 500。随着叶轮形状的变化，水流由离心泵的轴向进水、径向出水到混流泵的轴向进水、斜向出水，到轴流泵则变

为轴向进水、轴向出水。随着比转速的增加，水泵的性能也发生有规律的变化，由离心泵的小流量、高扬程，到轴流泵的大流量、低扬程。

关于比转速特性，还需注意以下情况：

（1）比转速系指单叶轮的参数，所以对于双吸泵（为双叶轮），取泵流量除以 2 以后的值。H 为水泵的扬程，m；对于多级泵，则取泵扬程除以级数后的数值。

（2）比转速不随转速的变化而改变，但其他工作参数不同时，比转速将不同。需要强调的是公式中的参数是额定参数，由此得出的才是该泵的比转速。一台泵只有一个比转速，即额定工况下的比转速。

（3）各个国家采用的单位不同，n_s 取的值也不相同。

（4）如果两台泵符合相似条件，它们的比转速必然是相等的；但如果两台水泵的比转速相等，则无法判断它们是否一定相似，因为几何形状并不一定相似。

Chapter 4　Pump Performance and Similar Conversion

The six operating parameters of the pump represent the performance of the pump. Each operating parameters has certain internal relations and change rules. If they are represented by a set of curves, the set of curves is called performance curves. Generally, when the pump rotating speed is at a certain value, the relationship between the head, shaft power, efficiency, vacuum suction height or NPSH with the discharge to draw various performance curves.

Only by mastering the performance of the pumps and being familiar with the characteristics of the performance curves of various pumps can we reasonably select and match the pumps in the pumping stations, correctly determine the installation elevation of the pumps and solve many problems encountered in the operation of the pump devices.

4.1　Theoretical performance curve

Because the flow condition of water body in a finite number of impellers is very complex, it is difficult to accurately calculate various losses. At present, the performance curve can not be determined by theoretical methods, and only some qualitative analysis can be done theoretically.

4.1.1　Relation curve between head and discharge

From the basic equation of the pump, the pump discharge is closely related to the head. In the velocity triangle

$$v_{u2} = u_2 - v_{m2}\cot\beta_2 , \quad v_{m2} = \frac{Q_T}{\pi D_2 b_2 \Psi_2}$$

Substituting the basic equation.

$$H_{T\infty} = \frac{u_2^2}{g} - \frac{u_2\cot\beta_2}{g\pi D_2 b_2 \Psi_2}Q_T \tag{4.1}$$

For a specific pump, π, D_2, b_2, Ψ_2 and β_2 are constant. When the rotating speed is fixed, u_2 is also a constant value. Eq. (4.1) indicates that the head changes with discharge is a linear equation. In centrifugal pumps, when the β_2 of the blade is less than 90 degrees, the theoretical head $H_{T\infty}$ decrease with the increase of Q_T. The straight line intersects the value $\frac{u_2^2}{g}$ on the axis with ordinate H, and the slope of the straight line depends on the value of β_2, as shown in Fig. 4. 1.

(1) When the number of blades is limited, the theoretical head is corrected by $H_T = KH_{T\infty}$ so that the relationship between $H_{T\infty}$ and Q_T is still a straight line, and its intersection on the ordinate H axis is the value $\dfrac{u_2^2}{gK}$.

(2) The actual head of the pump is equal to the theoretical head of the pump minus the hydraulic loss h in the pump, which h includes friction loss h_f and local loss h_j. The actual head curve of the pump is the head of curve Ⅱ minus h, namely curve Ⅲ $(H - Q_T)$.

(3) The volume loss of the pump is related to the pressure inside the pump. The larger the pressure, the greater the leakage. Its relationship can be expressed by the $H_T - q$ curve. When H_T is small (low pressure), there is no leakage in the pump. Even the water flow can pass through the intermittent overflow, which may make the discharge show a positive value. Since the actual leakage flow is reduced, the corresponding discharge minus the corresponding $H_T - Q$ on the curve $H - Q_T$ is worth the required $(H - Q)$ curve, that is the performance curve Ⅳ $(H - Q)$ of the theoretical head-discharge relationship of the pump.

4. 1. 2 Power-discharge curve

Shaft power is the sum of water power and mechanical loss power.

The theoretical input power N_T of the pump is

$$N_T = K\rho Q_T u_2 \left(u_2 - \frac{\cot\beta_2}{\pi D_2 b_2 \Psi_2} Q_T \right) \qquad (4. 2)$$

When β_2 is less than 90 degrees, curve Ⅰ $(N_T - Q_T)$ on Fig. 4. 2 is a downward curved parabola.

Mechanical loss of power is almost independent of discharge. Curve Ⅱ $(N_T - Q_T)$ is obtained by adding N_m to Curve Ⅰ. The final input power-discharge curve Ⅲ $(N - Q_T)$ is obtained by subtracting the leakage q from the horizontal coordinate of curve Ⅱ.

4. 1. 3 Efficiency-discharge curve

From the known $H - Q$ curve and $N - Q$ curve Ⅲ, the efficiency values at each corresponding discharge can be obtained.

$$\eta = \frac{\rho g Q H}{N_a}, \quad N_a > 0$$

When Q is 0, $\eta = 0$; When H is 0, $\eta = 0$, the efficiency-discharge curve is a parabola passing through a point on the coordinate origin and a point on the abscissa (Fig. 4. 2).

4. 2 Real performance of water pump

The theoretical performance curve can only be qualitatively analyzed, and it is difficult to determine quantitatively at present. In practice, the performance curve must be drawn by the measured method. The performance curves, performance tables and type spectra drawn in the pump samples books are all measured or converted according to the similarity of model tests.

4. 2. 1   Basic characteristics

At the rated speed of the pump or under the design placement angle of the blade, the $H-Q$, $N-Q$, $\eta-Q$, H_s-Q or $NPSH-Q$ curves drawn with Q as X-axis and H, N, η, H_s or $NPSH$ as Y-axis are called basic characteristics.

Fig. 4. 3, Fig. 4. 4 and Fig. 4. 5 show the basic performance curves of the three types of pumps.

From the basic performance curves of the above three types of vane pumps, their characteristics and differences can be seen.

1. Head-discharge curve ($H-Q$)

The head-discharge curves of the three pumps are all descending curves, but the descent of the centrifugal pump is gentle and that of the axial flow pump is steep. Most axial flow pumps have Hump Region when the discharge is 40%- 60% of the design condition, which is the hydraulic unstable area. The head-discharge curve of the mixed flow pump also declines gently, which is between the centrifugal pump and the axial flow pump.

2. Power-discharge curve ($N-Q$)

The power-discharge curve of the centrifugal pump is rising, as shown in Fig. 4. 3. At zero discharge, the power is much smaller than that at design condition. The power-discharge curve of the axial flow pump is opposite to that of the centrifugal pump, which has the characteristics of decline. At zero discharge, the power reaches the maximum value, which can reach about 2 times of the power under the design condition. The power-discharge curve of the mixed flow pump is flat, when the discharge changes, the change of power is small, even when the flow rate is zero, the power increase rate is small, so the mixed flow pump runs smoothly and has better characteristics.

From the characteristics of the power-discharge curve, it is known that the centrifugal pump should be started by closing the valve to make the power machine start smoothly at low power, and then gradually open the gate valve after starting. Axial flow pump should be started by opening the valve. In general, it is forbidden to install the gate valve on the outlet pipe of the axial flow pump, so as to avoid serious overload of power after closing valve to start pump, resulting in damage to the power machine. Starting of closed valve or open valve of mixed flow pump depends on starting mode.

3. Efficiency-discharge curve ($\eta-Q$)

The general trend of the efficiency-discharge curves of the three pumps is to decrease from the best efficiency point (BEP) to both sides, but they still have their own characteristics. The efficiency-discharge curves of centrifugal and mixed flow pumps change gently on both sides of the BEP and the high efficiency zone has a wide range, which reflects that the centrifugal and mixed flow pumps have a large range of discharge changes, and the efficiency changes little, which is conducive to discharge regulation. The efficiency-discharge curve of the axial flow pump drops steeply on both sides of the BEP and the range of the high efficiency zone is narrow.

4. Permitted vacuum height or critical $NPSH$ curve (H_s-Q, $NPSH-Q$)

Both of these curves represent the water absorption performance of the pump and the condition of cavitation, but the change rules of the two are quite different. The former is a descending curve, and the latter is a curve with the minimum extremum.

4.2.2　General characteristic curve

The general characteristic curve of pump is to draw the performance curve of the same pump at different rotating speeds or different blade angles in a certain way on a map.

Fig. 4.6 shows a general performance curve of a centrifugal pump at 900r/min, 1000r/min, 1100r/min and 1200r/min. Fig. 4.7 shows the general performance curve of the axial flow pump when the blade installation angle is $-6°$, $-4°$, $-2°$, $0°$, $+2°$, $+4°$. The downward inclined solid curve in the figure represents the head-discharge curve, and the contour line in the figure represents the isoelectric power line and the equivalent efficiency line.

Through the general characteristic of pump, other parameters corresponding to the pump at different rotating speeds (or different installation angles) and different discharges, such as head, power, efficiency and permitted vacuum suction height (or permitted $NPSH$), can be easily found out.

4.2.3　Performance table

In practical application, in order to find the performance of the BEP of the pump conveniently and quickly, so as to select the type, the performance parameters of the pump at a certain rotating speed, i.e. the performance table (Table 4.1), are commonly expressed in the form of tables.

4.2.4　Comprehensive type spectrum

The so-called comprehensive type spectrum of water pump is formed by drawing the areas enclosed by the head within the BEP and the efficiency areas of various pumps on the same graph, as shown in Fig. 4.8. It can be used to quickly and conveniently find pump types. With the application of modern information technology, a database of pump types has been established, which is more convenient and fast to search.

4.2.5　Pump system characteristic

The aforementioned performance curves, performance tables, comprehensive performance curves, type spectra, etc. are all pump performance, usually obtained from pump samples, however in actual pumping station engineering, it is necessary to obtain performance of pump system. The head of the pump system is the energy difference between the inlet and outlet of the device, and the head loss of the inlet and outlet pipes is less than the head of the water pump. Fig. 4. 9 shows the performance curve of the axial flow pump system.

4. 2. 6   Complete characteristics (four quadrant characteristics)

When the pump is run in normal operation, its head, discharge, rotating speed and power are all positive, but under abnormal circumstances, some or all operating parameters may appear negative.

The performance curve is divided into four quadrants, and the basic performance curve described above is the performance curve in the first quadrant. If the head and discharge of the pump are negative, the performance curve of the pump will exceed the first quadrant and run in the second, third and fourth quadrants to obtain the performance curve of the pump under all normal and abnormal conditions. The complete characteristics curve is the same as that of the pump in the first quadrant and can only be obtained by test.

Fig. 4. 10 shows the complete characteristics of the double suction pump. Complete characteristics curve is very useful to solve the problems of water hammer and run – off speed of pump.

4. 3   Pump similarity theory and application

The working medium of the pump is liquid. Its movement in the impeller and flow channel is affected by various forces and boundary conditions, showing a complex flow state. There are some difficulties in solving the motion law and working performance by theoretical methods alone, which need to be solved by experimental methods. Even if a new pump is designed by theoretical methods, it needs to be verified by model tests. Therefore, the experimental method is widely used in the research of hydraulic machinery. The theoretical basis of the experiment is the theory of hydraulic similarity. Applying the similarity theory to the study of water pumps can solve the following problems:

(1) The design and manufacture new pumps based on to the results of model tests.

(2) The performance conversion of two pumps with similar geometry.

(3) The performance conversion of the same pump at different rotating speeds.

Therefore, the similarity theory is not only used in the design and manufacture of water pumps, but also to solve problems in the operation of water pumps.

4. 3. 1   Similar conditions

If two pumps maintain similar flow patterns, the following three similar conditions must be met.

(1) Geometric similarity means that the flow passage, impeller shape and size of the two pumps should be similar, and their roughness should be similar. It is easy to ensure that the absolute values of roughness are equal in technology, but it is difficult to ensure

that the relative roughness (Δ/D) is equal. Therefore, in geometric similarity, the similarity of shape and size is the main factor, and then roughness similarity is considered.

(2) Kinematics similarity means that the velocity field of water flow is similar. The ratio of the velocity of each pair of the same name on the corresponding particle of water flow in two pumps is a constant, and the angle between the velocities is equal.

(3) Dynamic similarity refers to the force on the corresponding particles of the water flow in the two pumps, whose nature is proportional to the direction and magnitude. The forces acting on the water flow inside the pump include gravity, pressure, inertia and viscous forces. Among them, gravity similarity criterion—Froude numbers are equal; pressure similarity criterion—Euler numbers are equal; inertial force similarity criterion—Strouhal numbers are equal; viscous force similarity criterion—Reynolds numbers are equal.

4. 3. 2　Pump affinity law

Pump affinity law means that the discharge, head and shaft power of two pumps follow a certain regularity change relationship under the premise of geometrical similarity, Rinematics similarity and dynamic similarity.

1. First affinity law

The flow through the pump impeller is $Q=\pi Db\Psi v_m\eta_V$, then the flow ratio of prototype (P) and model (M) is

$$\frac{Q_P}{Q_M}=\frac{\pi D_P b_P \Psi_P v_{mP}\eta_{VP}}{\pi D_M b_M \Psi_M v_{mM}\eta_{VM}}$$

and

$$\frac{v_{mP}}{v_{mM}}=\frac{n_P D_P}{n_M D_M}, \frac{b_P}{b_M}=\frac{D_P}{D_M}$$

When the geometric dimensions and rotational speeds of the two pumps are not very different, let $\Psi_P=\Psi_M$, $\eta_{VP}=\eta_{VM}$, then

$$\frac{Q_P}{Q_M}=\left(\frac{D_P}{D_M}\right)^3 \frac{n_P}{n_M} \qquad (4.3)$$

2. Second affinity law

If the pump head is $H=\dfrac{u_2 v_{u2}}{g}\eta_h$, then the ratio of prototype to model head is

$$\frac{H_P}{H_M}=\frac{u_{2P}v_{u_2P}\eta_{hP}}{u_{2M}v_{u_2M}\eta_{hM}}$$

because

$$\frac{u_{2P}}{u_{2M}}=\frac{v_{u_2P}}{v_{u_2M}}=\frac{n_{1P}D_{1P}}{n_{1M}D_{1M}}$$

Let $\eta_{hP}=\eta_{hM}$, then

$$\frac{H_P}{H_M}=\left(\frac{D_P}{D_M}\right)^2\left(\frac{n_P}{n_M}\right)^2 \qquad (4.4)$$

3. Third affinity law

Pump shaft power is $N=\dfrac{\rho g QH}{\eta}$, then the ratio of prototype to model power is

$$\frac{N_P}{N_M}=\frac{\rho g Q_P H_P \eta_M}{\rho g Q_M H_M \eta_P}=\frac{Q_P H_P \eta_M}{Q_M H_M \eta_P}$$

Substitute Eq. (4.3) and Eq. (4.4) into the upper formula, and set $\eta_P=\eta_M$, then

$$\frac{N_P}{N_M}=\left(\frac{D_P}{D_M}\right)^5\left(\frac{n_P}{n_M}\right)^3 \tag{4.5}$$

The pump affinity law shows that if the two pumps are similar, their discharge ratio is proportional to the third power of the linear size ratio and the first power of the rotating speed ratio of the two pumps; the head ratio is proportional to the square of the linear size ratio and the square of the rotating speed ratio; and the ratio of shaft power is proportional to the fifth power of the linear size ratio and the third power of the rotating speed ratio.

The above deduction is based on the assumption that the efficiency remains unchanged. This assumption can only be established if the rotating speed and linear size change little. Therefore, the size and rotating speed of similar pumps are limited.

4.3.3 Proportional law of pump

For the same pump, because $\dfrac{D_P}{D_M}=1$, according to the affinity law of the pump, it can be concluded that

$$\frac{Q_1}{Q_2}=\frac{n_1}{n_2} \tag{4.6}$$

$$\frac{H_1}{H_2}=\left(\frac{n_1}{n_2}\right)^2 \tag{4.7}$$

$$\frac{N_1}{N_2}=\left(\frac{n_1}{n_2}\right)^3 \tag{4.8}$$

Where, the subscripts "1" and "2" respectively indicate the operating conditions of the pumps at different rotating speeds.

Proportional law formula shows that when the rotating speed of the pump changes, the discharge of the pump is proportional to the rotating speed, the head is proportional to the square of the rotating speed, and the shaft power is proportional to the third power of the rotating speed.

4.3.4 Performance conversion of pumps

According to the law of affinity or proportionality of the pump, the pump performance of different sizes and rotating speeds can be easily converted.

It must be noted that when deriving the pump affinity law, it is assumed that the efficiency of similar pumps is equal, but in practice the efficiency of two pumps are different. In addition, due to the "scale effect" caused by the inability to achieve complete mechanical similarity, such as gap, surface roughness, etc., the results of similar conversion are different from the actual situation. For the performance conversion of similar pumps with higher requirements, the above effects must be considered. There are provisions in acceptance specifications for model test of pumping stations in China, which can be referred to in relevant literatures.

4. 4   Specific speed

From the inertial force similarity criterion and pressure similarity criterion, a number of similarity criteria can be obtained:

$$n_s = \frac{1000}{60} \frac{n\sqrt{Q}}{(gH)^{3/4}} \qquad (4.9)$$

For customary reasons, g in Eq. (4.9) is removed and multiplied by 3.65 to change to a factorial Eq. (4.10), called specific speed.

$$n_s = \frac{3.65n\sqrt{Q}}{H^{3/4}} \qquad (4.10)$$

When the head of the model pump is 1m and the effective power is 1hp (1hp = 0.7355kW), its discharge can be calculated by the following formula:

$$Q_s = \frac{1000N_a}{\rho g H_s} = \frac{1000 \times 0.7355}{1000 \times 9.81 \times 1} = 0.075 \ (\text{m}^3/\text{s}) \qquad (4.11)$$

The rotating speed of this model pump is n_s, which can be obtained by substituting the third affinity law.

Adding a "ratio" in front of the rotational speed of the model pump is to take the meaning of comparison. If many different types and shapes of pumps are converted into their respective model pumps, and the relevant performance parameters. The head is 1m, the effective power is 1hp, and the flow rate is 0.075m³/s, the specific pump speed n_s can not be equal. It can therefore be used to compare and classify pumps. A lot of practical experience shows the relationship between pump type, geometry and operating performance and specific speed as shown in Table 4. 2.

From Table 4. 2, it can be seen that with the change of specific speed from small to large, the pump type changes regularly from centrifugal pump-mixed flow pump-axial flow pump. The specific speed of centrifugal pump is 50 – 300 and that of mixed flow pump is 300 – 500. In recent years, the specific speed of mixed flow pump has increased to about 600, while that of axial flow pump is more than 500. As the shape of the impeller changes, water flows from the centrifugal pump's axial inlet and radial outlet, the mixed flow pump's axial inlet and oblique outlet, and the axial flow pump becomes axial inlet and axial outlet. With the increase of specific speed, the performance of the pump also changes regularly, from small discharge and high head of centrifugal pump to large discharge and low head of axial flow pump.

The specific speed characteristics also need to be noted as follows:

(1) Because the specific speed is a parameter of a single impeller, for a double-suction pump, take the value of the pump flow divided by 2. The head (m) of the water pump is

H, for multi-stage pumps, take the head of the pump divided by the number of stages.

(2) The specific speed does not change with the change of speed, but it will vary with other operating parameters. It should be emphasized that the parameters in the formula are rated parameters, and the specific speed of the pump can be obtained. One pump has only one specific speed, which is the specific speed under rated conditions.

(3) Different countries use different units, but the values of n_s are different.

(4) If the two pumps meet similar conditions, their specific speeds must be equal, but if the specific speeds of the two pumps are equal, we may not be able to judge whether they must be similar, because the geometrical shapes are not necessarily similar.

第5章　水泵运行工况与调节

水泵在实际运行时的工况取决于水泵性能、管路水力损失以及上下游水位差的配合。这三种因素中任何一项发生变化，水泵运行工况都会随着变化。运行工况通常并不恰好与设计工况一致，有时甚至相差很多，以至于泵站整体效率偏低、长期偏工况运行、运行不经济。因此，应掌握泵站水泵运行工况的确定方法，以分析机组选型和水泵装置设计的合理性，同时也应掌握在已有水泵装置中运行工况的调节方法，以节约能源和合理运行水泵。

5.1　水　泵　工　况　点

水泵工况点是指水泵在确定的管路系统中，实际运行时所具有的扬程、流量以及相应的效率、功率等参数。已知水泵的流量是随扬程变化的，当其他条件一定时，在确定的扬程下对应一个确定的流量，这就是水泵的工况点。显然，这个工况点必定在扬程-流量曲线上，但在扬程-流量曲线上何处，还需根据进、出水位差（压差）和管路性能来确定。

5.1.1　管路水力损失及性能曲线

管路水力损失 h_l 由两部分组成：沿程水力损失 h_f 和局部水力损失 h_j，用公式表达为

$$h_l = h_f + h_j = \sum \lambda_i \frac{l_i}{d_i} \frac{v_i^2}{2g} + \sum \zeta_i \frac{v_i^2}{2g} \tag{5.1}$$

式中：λ 和 ζ 分别为沿程阻力系数和局部阻力系数；d 为圆管直径。

式中的流速 v 用流量 Q 来代替，则该式可以写为

$$h_l = \left(\sum \xi_{fi} l_i + \sum \zeta_i \frac{1}{2gA^2} \right) Q^2 \tag{5.2}$$

式中：A 为管道的过流断面面积；ξ_{fi} 为摩阻系数。

令 $\sum \xi_{fi} l_i + \sum \zeta_i \dfrac{1}{2gA^2} = S$（$S$ 为管路阻力系数），则有

$$h_l = SQ^2 \tag{5.3}$$

对于一个特定的管路，ξ_f、l、ζ、A 和 g 均是定值，所以 S 是一定值，管路水力损失 h_l 与流量的平方成正比，式（5.3）表示一条通过坐标原点的二次抛物线，这条抛物线称为管路水力损失曲线，如图 5.1 所示。

当泵站上下游水位确定，管路所需要的扬程 H_r 等于上下游水位差（也称净扬程 H_{st}）加上管路损失，即

$$H_r = H_{st} + h_l \tag{5.4}$$

或

$$H_r = H_{st} + SQ^2 \tag{5.5}$$

如按上式绘制水泵管路所需扬程随流量而变化的关系曲线，则得到图 5.2 所示的需要扬程曲线。它是一条起点为（$Q=0$，$H=H_{st}$）的二次抛物线。

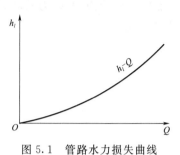

图 5.1 管路水力损失曲线

Fig. 5.1 Hydraulic loss curve of the pipeline

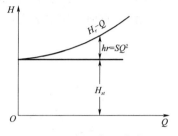

图 5.2 需要扬程曲线

Fig. 5.2 Required head curve

5.1.2 水泵工况点的确定

将水泵的性能曲线（H-Q）和需要扬程曲线 H_r-Q 绘在同一张坐标系中，则这两条曲线的交点 A 就是水泵的工况点（图 5.3）。

由图中可以找出 A 点的扬程 H_A、流量 Q_A、效率 η_A。在 A 点，水泵提供的扬程 H_A 和管路所需的扬程 H_r 相等，水泵抽送的流量等于管路所通过的流量。即工况点 A 是水泵在一定的抽水装置和一定的上下游水位条件下，达到能量和流量的平衡。这个平衡是有条件的和相对的，一旦水泵或管路中的一个或同时发生变化，平衡就被打破，并在新的条件下出现新的平衡。

由 H_A、Q_A 及 η_A 可以校核水泵装置是否在高效区运行。

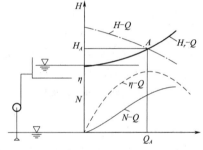

图 5.3 水泵工况点（图解法）

Fig. 5.3 Duty point by graphical method

5.2 水泵串联与并联运行

5.2.1 水泵串联运行

在管网中当一台水泵的扬程不够时，可以采用两台或多台水泵串联。在水泵串联工作时，经过每台水泵的流量相等，但串联总扬程为该流量下各台水泵的扬程之和。如图 5.4 所示，串联后的总性能曲线可以用"纵加法"汇出，即把同一 Q 值时的各台泵的扬程值相加，就得到水泵Ⅰ和水泵Ⅱ性能曲线总和。水泵串联总性能曲线与管道性能曲线的交点 A 即为串联运行时的工况点。由 A 点向下作垂线，可得出每台泵的工况点 B、C，若此时对应每台泵的高效率点，则串联水泵符合经济运行的要求。由此可见，串联运行各台水泵的高效区流量最好相等。

5.2.2 水泵并联运行

如果不同时段流量的变化较大，为节省管路材料，可以采用水泵并联运行的方式使多台水泵合用一条出水管道，称为水泵并联工作。并联后的总性能曲线可用横加法绘制，即把同一 H 值时的各台泵的 Q 值相加，就得到并联总性能曲线。以同型号同水位两台泵的并联工作为例（图 5.5），水泵并联总性能曲线与管道性能曲线的交点 A 即为并联运行时的工况点。由 A 点向左作水平线，可得出每台泵的工况点 M 点。

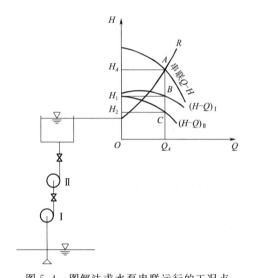

图 5.4　图解法求水泵串联运行的工况点

Fig. 5.4　Duty point of the pump in series by graphical method

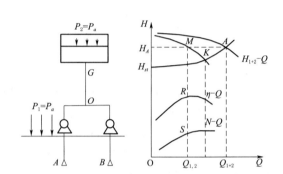

图 5.5　图解法求水泵并联运行的工况点

Fig. 5.5　Duty point of the pump in parallel by graphical method

由图 5.5 可知：

（1）采用等扬程下流量相叠加的方法，相当于把管道水力损失视为零的情况下求并联后的工况点。此等效水泵的流量，必须等于各台水泵在相同扬程时的流量之和。

（2）由于两台泵在同一吸水池中抽水，从吸水口 A、B 至联络管交汇点 O 的管径相同，长度也相等，故 $\sum h_{AO} = \sum h_{BO}$，$AO$ 与 BO 管中通过的流量均为 $Q/2$。每台水泵提水的需要扬程也相等。

（3）水泵并联工作时应注意：当两台性能曲线差别很大的水泵在一起工作往往是不合理的。

5.3　水泵工况的调节

在选择和使用水泵的实践中，常常会出现所确定的工况点偏离水泵设计工况点较远的情况，造成水泵装置效率降低、动力机过载及发生严重的汽蚀、振动等现象。因此必须采取改变管路性能曲线或改变水泵性能曲线的方法来移动工况点，使其满足要求。这种方法称为水泵工况的调节。常用的水泵工况调节方法有变速、变径和变角调节三种。

5.3.1　变径调节

沿外径车削离心泵或蜗壳式混流泵的叶轮，可以改变水泵性能曲线，用这种方法扩大水泵的使用范围，称为变径调节，也称车削调节。轴流泵不宜车削叶轮，否则需更换泵壳或加衬里。我国大多数离心泵，除了标准叶轮直径外，还有叶轮车削的两种以上变型，用字母"A""B""C"等表明。使用单位如有必要也可自行车削叶轮，来达到改变水泵工况点的目的。

1. 车削定律

叶轮直径车小后，与原来的叶轮并不保持几何相似，过流面积 $F_2 \neq F_{2a}$，出口安装角 $\beta_2 \neq \beta_{2a}$，相似条件受到破坏，所以不能用相似律来换算水泵的工作参数。但是当车削量不大时，认为过流面积和出口安装角在车削前后均相等，效率不变，这样车削前后的出水速度三角形可以认为是相似的，即水泵是运动相似的。

已知叶轮过水流量等于轴面速度乘以过水断面，即

$$Q = v_{m2} \pi D b_2 \Psi_2 \eta_V \qquad (5.6)$$

则车削前后流量之比为

$$\frac{Q}{Q_a} = \frac{v_{m2} \pi D b_2 \Psi_2 \eta_V}{v_{m2a} \pi D_a b_{2a} \Psi_{2a} \eta_{Va}} \qquad (5.7)$$

式中：下标 a 表示车削后的各参数。

在离心泵设计中，为了尽量减少叶槽内水流的脱壁现象，通常使叶槽的不同过水断面具有近似相等的面积，又根据前面的假设有 $\pi D b_2 \Psi_2 = \pi D_a b_{2a} \Psi_{2a}$，假定车削前后容积效率不变，$\eta_V = \eta_{Va}$，则上式可以简化为

$$\frac{Q}{Q_a} = \frac{v_{m2}}{v_{m2a}} \qquad (5.8)$$

假定车削前后的出口速度三角形相似，则有

$$\frac{v_{m2}}{v_{m2a}} = \frac{v_{u2}}{v_{u2a}} = \frac{u_2}{u_{2a}} = \frac{Dn}{D_a n_a} = \frac{D}{D_a} \qquad (5.9)$$

根据式（5.8）、式（5.9）和水泵基本方程，可得到下列适用于车削调节的工作参数换算公式，即车削定律。

$$\left. \begin{aligned} \frac{Q}{Q_a} &= \frac{D}{D_a} \\ \frac{H}{H_a} &= \left(\frac{D}{D_a}\right)^2 \\ \frac{N}{N_a} &= \left(\frac{D}{D_a}\right)^3 \end{aligned} \right\} \qquad (5.10)$$

由车削定律消去 D/D_a 得

$$\frac{H}{Q^2} = \frac{H_a}{Q_a^2} = K \ 或 \ H = KQ^2 \qquad (5.11)$$

式（5.11）表示的是在 H-Q 坐标系中顶点在坐标原点的二次抛物线族，通常称为车削抛物线。在水泵使用中，如需要的工况点 $A (Q_a, H_a)$ 位于水泵 H-Q 性能曲线的下面，可以利用车削抛物线来求所需的车削量，以便车削后的新的性能曲线经过 A 点。

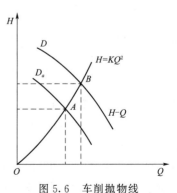

图 5.6 车削抛物线

Fig. 5.6 Cutting parabola

车削定律以出口速度三角形相似为前提，但车削前后的叶轮出口实际上并不完全相似，因此应用车削定律决定的 ΔD 后会产生误差，故必须对上述理论值按下式进行修正

$$\Delta D = K\ (D - D_a) \tag{5.12}$$

式中：K 是由实验得出的车削系数。

2. 车削量的范围及方式

（1）叶轮可车削量的大小与 n_s 有关，n_s 越高，允许车削量越小，比转速超过 350 的混流泵和所有的轴流泵不允许车削，否则容积损失过大，很不经济。

（2）车削量过大，会造成水泵水力效率、容积效率及机械效率较大幅度的下降。表 5.1 给出了允许的最大车削量与效率及车削量的关系。

表 5.1　　　　　　　　　　　　　允许的最大车削量

Table 5.1　　　　　　　　　　　Maximum allowable cutting value

比转速 specific speed	60	120	200	300	350	>350
许可最大车削量 permitted maximum cutting value	20%	15%	11%	9%	7%	0
效率下降值 efficiency decline value	每车削 10%下降 1% decrease by 1% every 10% of turning		每车削 4%下降 1% decrease by 1% every 4% of turning			

（3）车削方式。不同 n_s 的叶轮采用不同的车削方式，低比转速的叶轮，可以平车，即前后盖板同时车削；中、高比转速的离心泵，可车成倾斜的外圆，内缘直径大于外缘直径；混流泵的叶轮只车外缘，不车轮毂。

5.3.2　变速调节——比例律应用

变速调节就是用改变水泵转速的方法来改变水泵运行工况。水泵转速的改变，可以通过改变动力机转速或改变传动机构的转速比等方法来实现。

1. 相似工况抛物线

由水泵比例律，可以得到

$$\left(\frac{Q_1}{Q_2}\right)^2 = \frac{H_1}{H_2} = \left(\frac{n_1}{n_2}\right)^2 \tag{5.13}$$

故有

$$\frac{H_1}{Q_1^2} = \frac{H_2}{Q_2^2} = \frac{H}{Q^2} = K$$

所以

$$H = KQ^2 \tag{5.14}$$

式（5.14）是顶点在坐标原点的二次抛物线族，称为相似工况抛物线（图 5.7），K 为相应的抛物线常数。由于式（5.9）是从比例律推导而得的，所以符合 $H = KQ^2$ 的所有点都是相似工况点。在推导比例律时认为各点效率相等，所以这条曲线又是等效率曲线，A、B、C 等点的效率也是相等的。

2. 应用实例

在图 5.8 中，已知某泵在转速 n_A 时的曲线 H-Q，但所需的工况点 B（Q_B，H_B）并不位于该曲线上，为了使水泵能在新的工况点工作，求它所需的转速 n_B。

为此，可以将 Q_B、H_B 代入相似工况抛物线公式（5.14），求出 K 值，然后按照该式

绘出相似工况抛物线，和转速 n_A 时的曲线 H-Q 相交于 A 点，查出 A 点的参数 Q_A，H_A。工况点 A（Q_A，H_A）与工况点 B（Q_B，H_B）工况相似，通过比例律公式可求得 n_B。

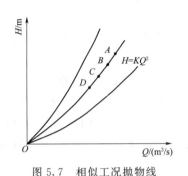

图 5.7　相似工况抛物线

Fig. 5. 7　Parabola for similar conditions

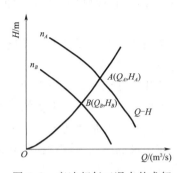

图 5.8　变速相似工况点的求解

Fig. 5. 8　Solution to similar duty point of variable speed regulation

3. 变速调节相关问题

（1）相似工况抛物线公式与车削抛物线公式在形式上是一样的，但两者是根据不同角度推导的，而基本原理都是相似理论的应用。对于不在水泵性能曲线上的工况点，既可用车削的办法得到，也可通过改变转速的办法得到。在采用变速调节时，还要注意不要使变速后的转速接近水泵的临界转速以防止引起共振而损坏机组。

（2）水泵的变速调节是有限制的，应在一定的范围内。若转速降幅过大，相似工况抛物线和等效率曲线不重合，降速后实际效率下降。此外，必需汽蚀余量与转速的二次方成正比增加，因此一般不宜采用增速的方法。

5.3.3　变角调节

改变叶片的安装角，使水泵性能曲线改变的方法称为水泵工况的变角调节，它适用于叶片安放角可以转动的轴流泵及导叶式混流泵。性能发生变化，达到调节目的，这就是变角调节。

1. 叶片安放角

叶片安放角指叶片外缘断面的弦线与叶片圆周速度之间的夹角。它有两种表示方法：一种是取安放角的实际角度；另一种是以叶片设计安放角度为 $0°$，大于设计值的角度为正值，小于设计值的为负值。通常在叶片的根部和轮毂面刻有按角度分的线条，如 $0°$、$-2°$、$+2°$、$-4°$、$+4°$ 等。

2. 变角调节原理

由图 5.9 可以看出：当安装角为 β_2 时，出口速度三角形为实线所示；安放角变为 β'_2（$\beta'_2 > \beta_2$）时，由于该点的圆周速度不变（均为 u_2），流量不变（v_{m2} 相同），由于 w_2 的方向改变了，速度三角形随之变为虚线所示。比较两三角形中的 v_{u2}，后者明显增大，根据基本方程 $H_{T\infty} = \dfrac{1}{g} u_2 v_{u2}$，可见 H 增加了，即在流量 Q 不变的情况下扬程 H 增加。所以

$H\text{-}Q$ 曲线上移，而这时效率变化很小。

3. 叶片角度调节方式

（1）半调节。在中小型水泵上，叶片用紧固螺栓固定在轮毂上，叶片和轮毂上刻有指示线和角度线，调节时松开螺母，即可转动叶片，一般在停机检修时进行，故不能实时调节。

（2）全调节。在大型泵上，通过液压系统或机械调节机构进行实时角度调节。

4. 水泵叶片全调节的优点

（1）叶片角度调节能在较大的流量范围内保持水泵运行效率基本不变。

（2）根据需要可以使动力机始终在满载的情况下运行，如图 5.10 所示，由于外河水位的变化，装置工况是变化的，用 Ⅰ、Ⅱ、Ⅲ 分别表示进水池最高水位、设计水位和最低水位，对应于最低扬程、设计扬程和最高扬程时的装置性能曲线。当叶片安装角为 0°时，其工况点分别为 A、C、B 三点。

从实际运行管理来说，希望动力机满负荷高效率运行。在高扬程时将角度调到 $-2°$，减小流量；在低扬程时将角度调到 $+2°$，增加流量。水泵运行的工况点分别在 D、E 点上，使动力机的轴功率保持在额定值左右满负荷运行。

（3）当水泵小角度启动时，阻力矩小，便于机组启动。在水泵启动前先将叶片角度调小，待启动完成后再将叶片角度调为正常。

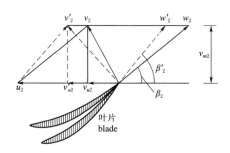

图 5.9　轴流泵变角调节原理

Fig. 5.9　Variable angle adjustment method of the axial flow pump

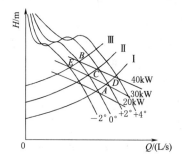

图 5.10　轴流泵变角调节的应用

Fig. 5.10　Application of variable angle adjustment of axial flow pump

Chapter 5   Operating Conditions and Regulation of Pumps

The actual operation of the pump depends on the performance of the pump, the hydraulic loss of the pipeline and the cooperation of the upstream and downstream water level difference. If any of these three factors changes, the operating condition of the pump will change with the factors changes. The operating conditions are often not exactly the same as the design conditions, sometimes quite different, so that the overall efficiency of the pumping station is low, long-term operation under partial conditions, and operation is uneconomical. Therefore, it is necessary to master the determination method of pump operating condition in pumping station to analyze the rationality of unit selection and pump device design, as well as the adjustment method of operating condition in existing pump devices, so as to save energy and operate the pump reasonably.

5.1   Duty point (operating point) of pump

Duty point of pump refers to the head, discharge, efficiency, power and other parameters of the pump in the determined pipeline system during actual operation. It is known that the discharge of the water pump varies with the head, when other conditions certainly, it corresponds to a certain discharge under a certain head, which is the duty point of the water pump. Obviously, this duty point must be on the head-discharge curve. The location on the head-discharge curve needs to be determined according to the water level difference (differential pressure) between inlet and outlet and the pipeline performance.

5.1.1   Hydraulic loss and performance curve of pipeline

The hydraulic loss h_l of a pipeline consists of two parts: the hydraulic loss h_f along the pipeline and the local hydraulic loss h_j, which are expressed by an equation:

$$h_l = h_f + h_j = \sum \lambda_i \frac{l_i}{d_i} \frac{v_i^2}{2g} + \sum \zeta_i \frac{v_i^2}{2g} \tag{5.1}$$

Where, λ and ζ are resistance coefficients along the way and local resistance coefficients respectively; and d is the diameter of the circular pipe.

Where the velocity v is replaced by the discharge Q, the equation can be written as:

$$h_l = \left(\sum \xi_{fi} l_i + \sum \zeta_i \frac{1}{2gA^2} \right) Q^2 \tag{5.2}$$

Where, A is the area of flow section of pipeline; ξ_{fi} is the coefficient of friction.

If $\sum \xi_{fi} l_i + \sum \zeta_i \dfrac{1}{2gA^2} = S$ (S is the resistance factor of the pipeline), then

$$h_l = SQ^2 \tag{5.3}$$

For a specific pipeline, ξ_f, l, ζ, A and g are fixed values, so S is a certain value. The hydraulic loss h_l of the pipeline is proportional to the square of the flow. Eq. (5.3) represents a quadratic parabola passing through the origin of coordinates. This parabola is called the hydraulic loss curve of the pipeline. See Fig. 5.1.

When the upstream and downstream water levels of the pumping station are determined, the required head H_r of the pipeline equals the difference between the upstream and downstream water levels (also known as net head H_{st}) plus the loss of the pipeline. That is

$$H_r = H_{st} + h_l \tag{5.4}$$

or

$$H_r = H_{st} + SQ^2 \tag{5.5}$$

If the relation curve of the required head of the pump pipeline with the flow rate is drawn according to the formula above, the required head – discharge curve shown in Fig. 5.2 can be obtained. It is a quadratic parabola with a starting point of ($Q=0$, $H = H_{st}$).

5.1.2 Determination of duty point

If the performance curve ($H - Q$) and the required head curve ($H_r - Q$) of the pump are plotted in the same coordinate system, then the intersection point A of the two curves is the duty point of the pump (see Fig. 5.3).

The head H_A, discharge Q_A and efficiency η_A of point A can be found from the figure. The head H_A provided by the A-point pump is equal to the head H_r required by the pipeline, and the discharge pumped by the pump is equal to the discharge through the pipeline. That is, duty point A is the balance of energy and discharge achieved by the pump under certain pumping system and certain upstream and downstream water level conditions. This balance is conditional and relative, and once one or both of the pumps or pipes changes, the balance is broken and new equilibrium occurs under new conditions.

H_A, Q_A and η_A can be check whether the pump unit is operating in BEP (Best Efficiency Point).

5.2 Duty point of pump in series and parallel

5.2.1 Duty point of pump in series

When the head of one water pump is insufficient in the pipeline network, two or more water pumps can be used in series. When the pump works in series, the discharge through each pump is equal, but the total head in series is the sum of the head of each pump under this discharge. As shown in Fig. 5.4, the total performance curve after series connection can be summarized by longitudinal addition, that is, the sum of the pump I and pump II performance curves can be obtained by summing the head values of each pump at the same Q value. The intersection point A between the total performance curve of the pump in series and the pipeline performance curve is the duty point of the pump in series operation. By making a vertical line downward from point A, the working conditions B and C of each pump can be obtained. If the corresponding BEP of each pump at this time, the pumps in series is in line meet the requirements of economic operation. It can be seen that the flow rate in the BEP of running each pump in series is best equal.

5.2.2 Duty point of pump in parallel

If the discharge varies greatly at different times, in order to save pipeline materials, multiple pumps can be operated in parallel to use one outlet pipe, which is called parallel operation of pumps. The total performance curve after parallel connection can be drawn by cross-addition method, i.e. the total performance curve in parallel can be obtained by adding up the Q values of each pump at the same H value. The parallel operation of two pumps with the same model and water level is taken as an example (Fig. 5.5). The point at which the total performance curve of the pump in parallel intersects with the pipeline performance curve A is the operating point in parallel operation. The operating point M of each pump can be obtained by making a horizontal line from point A to the left.

It can be seen from the Fig. 5.5.

(1) The method of superposition of discharge at equal head is equivalent to calculating the duty point after parallel connection when the hydraulic loss of pipeline is regarded as zero. The discharge of this equivalent pump must be equal to the sum of the discharge of each pump at the same head.

(2) Because two pumps are pumped in the same sump, and the diameter and length from suction port A and B to junction point O of the connecting pipe are the same, so $\sum h_{AO} = \sum h_{BO}$, the discharge through the AO and BO tubes is $Q/2$. The head required to lift water from each pump is also equal.

(3) When working in parallel, it should be noted that it is often unreasonable for two pumps with different performance curves to work together.

5.3 Regulation of pump duty point

In the practice of selecting and using pumps, the determined working conditions often deviate from the design duty point of pumps, resulting in the reduction of pump device efficiency, overload of power engines, and the occurrence of serious cavitation, vibration and other phenomena. Therefore, it is necessary to change the pipeline performance curve or pump performance curve to move the duty point to meet the requirements. This method is called the regulation of pump operating conditions. The commonly used adjusting methods of pump working conditions include variable speed regulation, cutting regulation and variable angle regulation.

5.3.1　Cutting regulation

Cutting the impeller of a centrifugal pump or volute-type mixed flow pump along the outer diameter can change the performance curve of the pump. This method expands the scope of use of the pump, known as variable diameter adjustment, also known as cutting regulation. Axial flow pumps should not cut impellers, otherwise, the pump casing needs to be replaced or lined. In addition to the standard impeller diameter, most centrifugal pumps in China also have two or more variants of impeller turning, which are indicated by the letters "A", "B", "C", etc. If necessary, the user can also cut the impeller by himself to achieve the purpose of changing the working point of the pump.

1. Cutting law

After cutting the impeller diameter, it does not keep geometric similarity with the original impeller, the overflow area F_2 is not equal to F_{2a}, the outlet installation angle β_2 is not equal to β_{2a}, and the similar conditions changed, so the working parameters of the pump cannot be converted by the similarity law. However, when the cutting volume is not large, it is considered that the overflow area and outlet installation angle are equal before and after cutting, and the efficiency is unchanged, so that the outlet velocity triangles before and after cutting can be considered similar, that is, the pump is similar in motion.

It is known that the impeller discharge is equal to the axial velocity multiplied by area of the cross-section, i. e.

$$Q = v_{m2} \pi D b_2 \Psi_2 \eta_V \tag{5.6}$$

Then the ratio of discharge before and after cutting is

$$\frac{Q}{Q_a} = \frac{v_{m2} \pi D b_2 \Psi_2 \eta_V}{v_{m2a} \pi D_a b_{2a} \Psi_{2a} \eta_{Va}} \tag{5.7}$$

Where, subscript a denotes the parameters after cutting.

In the design of centrifugal pump, in order to minimize the detachment of flow in the vane groove, the different water-passing sections of the vane groove usually have approximately equal area, and according to the previous assumption, $\pi D b_2 \Psi_2 = \pi D_a b_{2a} \Psi_{2a}$, assuming that the volume efficiency is unchanged before and after cutting, $\eta_V = \eta_{Va}$, the formula can be simplified as following

$$\frac{Q}{Q_a} = \frac{v_{m2}}{v_{m2a}} \tag{5.8}$$

Assuming that the outlet velocity triangles are similar before and after cutting, there are

$$\frac{v_{m2}}{v_{m2a}} = \frac{v_{u2}}{v_{u2a}} = \frac{u_2}{u_{2a}} = \frac{Dn}{D_a n_a} = \frac{D}{D_a} \tag{5.9}$$

According to Eq. (5. 8), Eq. (5. 9) and basic equation of pump, obtain the following working parameter conversion formula suitable for turning adjustment, which is Law of Turning.

$$\left. \begin{array}{l} \dfrac{Q}{Q_a}=\dfrac{D}{D_a} \\[2mm] \dfrac{H}{H_a}=\left(\dfrac{D}{D_a}\right)^2 \\[2mm] \dfrac{N}{N_a}=\left(\dfrac{D}{D_a}\right)^3 \end{array} \right\} \qquad (5. 10)$$

By eliminating D/D_a from the law of cutting:

$$\dfrac{H}{Q^2}=\dfrac{H_a}{Q_a^2}=K \text{ or } H=KQ^2 \qquad (5. 11)$$

This equation represents a family of quadratic parabola whose vertex are at the origin of coordinates in the $H-Q$ coordinate system, commonly referred to as cutting parabola. In the use of pumps, if the required duty point A $(Q_a,\ H_a)$ is located below the pump $H-Q$ performance curve, cutting parabola can be used to calculate the required cutting a-mount, so that the new performance curve after cutting passes through point A.

The cutting law is based on the similarity of outlet velocity triangles, but the impeller outlets before and after cutting are not exactly similar. Therefore, errors will occur after applying ΔD determined by cutting law, so the above theoretical values must be corrected according to the following formula

$$\Delta D=K\ (D-D_a) \qquad (5. 12)$$

Where, K is the cutting coefficient obtained from experiments.

2. Scope and mode of cutting quantity

(1) The cutable amount of impeller is related to n_s. The higher n_s is, the smaller the allowable cutting amount is. Mixed flow pumps with specific speed over 350 and all axial flow pumps are not allowed to turn. Otherwise, the volume loss is too large, which is very uneconomical.

(2) Too much cutting will cause a large reduction in hydraulic efficiency, volumetric efficiency and mechanical efficiency of the pump. Table 5. 1 shows the relationship between the maximum allowable cutting amount, efficiency and cutting amount.

(3) Cutting method. The impellers of different n_s adopt different cutting methods. The impellers with low specific speed can be cuted flat, i. e. front and rear cover plates at the same time. The centrifugal pumps with medium and high specific speeds can be cuted into inclined outer circles with inner edge diameter larger than outer edge diameter. The impellers of mixed flow pumps only cut the outer edge without cutting the hub.

5. 3. 2　Variable speed regulation—Application of proportional law

Variable speed regulation is to change the operating condition of the pump by changing the rotating speed of the pump. The change of pump speed can be achieved by changing the speed of power engine or the speed ratio of transmission mechanism.

1. Parabola under similar working conditions

From the proportional law of the pump, it can be obtained

$$\left(\frac{Q_1}{Q_2}\right)^2 = \frac{H_1}{H_2} = \left(\frac{n_1}{n_2}\right)^2 \tag{5.13}$$

So there is

$$\frac{H_1}{Q_1^2} = \frac{H_2}{Q_2^2} = \frac{H}{Q^2} = K$$

as a result

$$H = KQ^2 \tag{5.14}$$

Eq. (5.14) is a family of quadratic parabola with vertex at the origin of coordinates, called parabola under similar conditions (Fig. 5.7), and K is the corresponding parabolic constant. Since the Eq. (5.9) is derived from the proportional law, all points conforming to $H = KQ^2$ are similar working conditions. When deducing the proportional law, it is considered that the efficiency of each point is equal, so this curve is an equivalent efficiency curve, and the efficiency of points A, B, C is equal.

2. Application examples

In Fig. 5.8, the curve $H - Q$ of a pump at rotating speed n_A is known, but the required duty B (Q_B, H_B) is not located on the curve. In order to make the pump work at B point, the required speed n_B is required.

For this purpose, Q_B and H_B can be substituted into parabolic Eq. (5.14) of similar working conditions to obtain K value, and then parabolic curve of similar working conditions can be drawn according to this formula. Curve $H - Q$ intersects point A when speed n_A, and parameter Q_A and H_A of point A can be found. Point A (Q_A, H_A) is similar to duty point B (Q_B, H_B), and n_B is obtained by the formula of proportional law.

3. Related issues

(1) The parabolic formulas under similar conditions are the same in form as those for cutting, but both are derived from different angles, and the basic principles are the application of similar theory. For the duty point not on the pump performance curve, it can be obtained either by cutting or by changing the speed. When using variable speed adjustment, it is also necessary to pay attention not to make the speed after variable speed close to the critical speed of the pump to prevent resonance and damage the unit.

(2) Variable speed regulation of pump is limited and should be within a certain range. If the speed drop is too large, the parabolic curve and equivalent efficiency curve under similar conditions do not coincide, and the actual efficiency decreases after speed reduction. In addition, the $NPSH_r$ is proportional to the quadratic of the speed, so the method of increasing the speed is generally not suitable.

5.3.3　Variable angle regulation

The method of variable angle adjustment of the pump to change the performance curve of the pump is called the variable angle adjustment of the pump duty point. It is suitable for axial flow pump and guide vane mixed flow pump with blades changing. The performance changes to achieve the purpose of adjustment, which is variable angle adjustment.

1. Blade setting angle

The blade setting angle refers to the angle between the chord of the outer edge section of the blade and the circumferential velocity of the blade. It has two representations: one is the actual angle of the setting angle; the other is the vane design setting angle of $0°$, the angle greater than the design value is positive, and the angle smaller than the design value is negative. In general, the blade base and hub face are engraved with angular lines, such as $0°$, $-2°$, $+2°$, $-4°$, $+4°$, etc.

2. Principle of variable angle adjustment

It can be seen from Fig. 5. 9 that when the installation angle is β_2, the outlet velocity triangle is shown as a solid line; when the installation angle becomes β'_2 ($\beta'_2 > \beta_2$), because the circumferential velocity of this point is unchanged (both are u_2), the discharge is unchanged (same as v_{m2}), and because the direction of w_2 changes, the velocity triangle is then shown as a dotted line. Comparing v_{u2} in the two triangles, the latter increases obviously. According to the basic equation $H_{T\infty} = \dfrac{1}{g}u_2 v_{u2}$. H increases, i. e. the head increases when the discharge Q is unchanged. So the $H-Q$ curve moves up, and the efficiency changes little.

3. Regulation method of adjustable blade

(1) Semi-regulation. On small and medium-sized vane pumps, the blades are fixed on the hub with fastening bolts, and the blades and the hub are engraved with indicator lines and angle lines. When adjusting, the blades can be rotated by loosening the nuts. Generally, the blades are carried out during shutdown maintenance, so real-time adjustment cannot be achieved.

(2) Full regulation. On large-sized pumps, real-time angle adjustment is performed by hydraulic system or mechanical adjusting mechanism.

4. Advantages of full regulation of pump blades

(1) The blade angle adjustment can keep the pump operation efficiency basically unchanged in a larger discharge range.

(2) According to the need, the power machine can always operate under full load, as shown in Fig. 5. 10. Due to the change of water level in the outer river, the duty point of the system is changed. I, II and III are used to represent the maximum water level, the design water level and the minimum water level of the sump, respectively, corresponding to the performance curve of the system at the lowest head, the design head and the maximum head. When the blade installation angle is $0°$, the duty points are A, C and B.

In terms of actual operation management, it is hoped that the power engine will operate at full load and with high efficiency. Adjust the angle to $-2°$ at high head to reduce the discharge; at low head to $+2°$ to increase the discharge. The duty point of the pump are at D and E points respectively, so that the shaft power of the power machine is kept at full load around the rated value.

(3) When the pump starts running at a small angle, the resistance moment is small, which is convenient for the pump unit to start. Turn blade angle down before starting the pump, and then adjust the blade angle to normal after starting.

第6章 水泵中的空化与汽蚀

空化与汽蚀又称空穴与空蚀,是水力机械、水工建筑物、船舶推进器等过流部件中的异常现象。1874年,雷诺第一次从理论上阐述并在实验室中观察空化现象,捕捉到了温水中蒸汽—气泡的运动规律。1895年,在一次对横渡大西洋的油轮的螺旋桨进行检查时,首次发现海洋轮船的螺旋桨表面布满了小孔,严重的被蚀成海绵状,甚至穿孔,更为严重时甚至导致整个叶片脱落。1897年巴纳比和帕森斯在"果敢号"鱼雷艇和几艘蒸汽机船相继发生推进器效率严重下降事件以后,提出了"空化"的概念,并指出在液体和物体间存在高速相对运动的场合就可能出现空化。人们从此开始认识水的空化和实际的汽蚀。再后来,人们不但在螺旋桨上,在轴流泵、离心泵、水轮机、深水武器甚至管道中都发现了这种的现象。

水泵运行时发生空化与汽蚀,会使水泵性能恶化直至断流,或使叶轮及泵壳等过流部件产生蜂窝麻面而被破坏。水泵汽蚀性能同时也是确定水泵安装高程的决定性因素。因此,对于水泵中的空化与汽蚀问题,必须予以高度重视。

6.1 空 化 机 理

6.1.1 水的汽化与空化

水在一定的压力(指压强)下,温度升高到某一临界值时,便发生汽化而出现气泡,即在水体中出现气相,称为沸腾,此临界温度称为沸点。其他液体也会产生同样的现象,只是沸点有所不同。

与沸腾相反,对于某种液体,当其温度一定,而液体内压力降低到某一临界值时,液体中也将出现汽化,产生无数充满蒸汽和气体的气泡,液体的连续性遭到破坏,这种因压力降低而形成气相的过程,称为空化现象。相应于汽化的临界压力,称为饱和蒸汽压力或汽化压力。不同的水温有不同的汽化压力,表6.1列出了不同水温时的汽化压力值,用水柱高表示,单位为 m。

表 6.1 不同水温时的汽化压力

Table 6.1 Vaporization pressure at different water temperatures

水温 water temperature/℃	20	30	40	50	60	70	80	90	100
汽化压力 vaporization pressure $(P_{cav}/\rho g)$ /m	0.24	0.43	0.75	1.25	2.02	3.17	4.82	7.14	10.33

沸腾与空化这两种现象,从物理本质上看,至少在它们的初生问题上是一致的,只是液体在形成汽化的原因上有所不同,沸腾是升温而引起的,空化则是压降而引起的。空化

的发生条件一般可归结为

$$P \leqslant P_{cav} \tag{6.1}$$

上述因液体中压力降低而产生空化的解释，常被工程界所采用。在水力机械中，水流压力降低往往是流速过大造成的，因此常采用一个由压力、流速组合的无因次数来表示水流发生空化可能的程度，称为水流的空化数，其表达式为

$$\sigma = \frac{p_\infty - p_v}{\frac{1}{2}\rho v_\infty^2} \tag{6.2}$$

式中：p_∞、v_∞ 分别为水流中未受干扰处的压力和流速；p_v 为水流中某点的压力。

当压力 p_v 降低到临界压力 P_{cav} 时，则发生空化。空化初生时的 σ 值可称为初生空化数 σ_i，一般而言，空化数越大表示越不易发生空化，对于水泵叶片，$\sigma_i = 0.1 \sim 0.2$。

6.1.2　汽蚀过程

如果过流部位的局部区域的绝对压力下降到当时温度下的汽化压力时，水体便在该处开始汽化，产生蒸汽，形成气泡，这些气泡随水流向前流动，至高压处时，气泡周围的高压液体致使气泡急骤缩小以至破裂（凝结），在气泡凝结的同时，液体质点将以高速冲击空穴中心，这些质点互相撞击而产生局部高压，同时伴有爆裂的声响。如果气泡破灭的位置恰好在固体表面处，这种冲击就会直接作用于固体表面，由于接触面积极小，压强值极大，加之空泡溃灭的频率极高，每秒可达数万次，对固体表层冲击极大，致使固体表面发生疲劳剥落，出现蜂窝状的凹坑，进而脱落，严重时甚至击穿，导致固体边壁破坏。这一现象就称为"汽蚀"。

空化与汽蚀作为十分复杂的物理化学现象，虽然这两种现象的产生是紧密相连的，但两者的宏观特点及机理不同，主要区别如下：

（1）空化是流体动力学现象，是无数微气泡发育、膨胀和溃灭的过程，而汽蚀是气泡溃灭时对边壁材料产生机械破坏、化学腐蚀以及电化作用等更为复杂的过程。

（2）空化产生于低压区（小于汽化压力的空间），汽蚀发生在高压区（压力回升空间）。

（3）汽蚀的起始点通常在空化的起始点之后，有时还相隔较远。

（4）由上述可进一步推论，对具体边界而言，有空化不一定必有汽蚀，如对超空化泵的叶轮而言，在某种程度上可以说避免了汽蚀。

（5）在同一空化形态下，固体边界条件及材料性质不同，汽蚀的强度可能不同。

6.2　诱发泵汽蚀的外界水力因素

在水泵运行时局部区域的液体压力低于工作温度下的汽化压力，是发生空化、进而导致汽蚀的条件。在水泵装置中造成压力过低的原因，除水泵本身结构及性能因素外，也有外界水力因素方面的原因。

1. 吸水装置压力过低

（1）水泵吸水高度过大，使水泵进口的真空度增加，绝对压力下降。

（2）安装水泵地点海拔较高，大气压力较低，引起吸入装置压力下降。

（3）水泵流量增加，流速加大，引起吸入装置压力下降。

（4）被抽吸的液体温度较高，汽化压力较大，容易汽化。

2. 泵内局部区域压力过低

（1）当泵运行偏离设计工况，进入叶轮内的水流发生撞击、漩涡等现象使局部流速加大，局部压力降低。

当水泵运行在大于设计流量工况时，进口相对速度方向偏向叶轮旋转方向，β_1 角增大，叶片前缘正面发生漩涡，容易产生负压而汽化。当水泵运行在小于设计流量工况时，进口相对速度方向偏向反转方向，β_1 角减小，叶片背面发生漩涡，加重了叶片背面低压区汽化的程度，如图 6.1 所示。

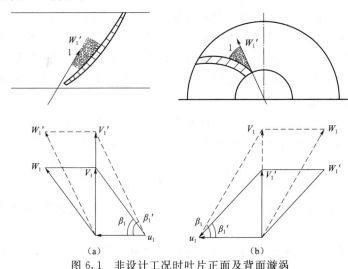

图 6.1　非设计工况时叶片正面及背面漩涡

Fig. 6.1　Vortexes on the front and back of blades under undesigned conditions

（a）叶片正面；（b）叶片背面

(a) Front surface of blade；(b) Back surface of blade

（2）采用肘形或斜式进水流道的轴流泵，由于进水流道弯道内侧易脱流（图 6.2），造成水泵叶轮进口流速分布不均匀。当叶片旋转到弯道外侧时，叶片前缘正面形成负压区；当叶片旋转到内侧时，叶片背面形成负压区。因此如果进水条件不良，进水流道出口流速分布不均，就有使叶片正面及反面同时出现负压而发生汽蚀的可能。

（3）泵内水流经过狭窄的间隙部位时，如轴流泵叶轮外缘与泵壳的间隙，离心泵蜗壳隔舌等部位，由于流速很大，间隙处压力下降而达到汽化条件。

（4）泵过流部分表面铸造质量较差，某些粗糙凸出物后面产生局部漩涡，引起局部压力下降。

（5）由于机组振动，在水中的某些部件发生某种频率的振动，在部件两侧正、负压交替产生，当负压达到汽化压力时即产生振动型气穴。

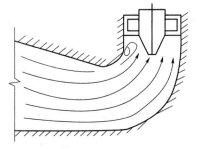

图 6.2　进水流道弯道内侧脱流示意图

Fig. 6.2　Flow separation in the bend of inlet passage

6.3　水泵中汽蚀的类型

在水泵汽蚀试验中，采用闪频仪透过观察窗可以清楚地观测到水泵叶轮内发生汽蚀（空化）的情况，用高速摄影机还可以拍摄到汽蚀（空化）发生的过程。试验表明，根据汽蚀形态分为游移型汽蚀、固定型汽蚀、漩涡型汽蚀及振动型汽蚀等。根据泵内汽蚀发生的部位，大致可分为翼型汽蚀、间隙汽蚀及流道中的漩涡型汽蚀三种类型。

1. 翼型汽蚀

由于叶片翼型型式而发生在叶片正、反面的汽蚀称为翼型汽蚀。产生汽蚀原因是水泵在非设计工况下运行时，叶片进口相对速度方向偏离设计方向，引起叶片正面或反面压力降低。

随着水泵工况的变化，翼型汽蚀的类型发生变化，可分为以下三种类型：

（1）Ⅰ类汽蚀。Ⅰ类汽蚀发生在叶轮外缘叶片进口背面处，此时流量小于最高效率点流量，当流量继续减小，扬程继续增高，叶片进口冲角增大，叶片背面的汽蚀区域逐步向出口边和叶片根部扩展。

（2）Ⅱ类汽蚀。Ⅱ类汽蚀发生在叶轮外缘、根部、叶片进口正面处，此时流量大于最高效率点流量，当流量继续增大，扬程继续降低，叶片进口冲角减小，叶片正面的汽蚀区域向中部扩展。

（3）Ⅲ类汽蚀。Ⅲ类汽蚀发生在叶片外缘和根部，从叶片近前端向后达叶片大部分区域，此时叶片角度较大，流量也大。由于汽蚀发生区域大，常导致水泵性能下降，被称为"有害汽蚀"。

2. 间隙汽蚀

间隙汽蚀主要发生在水流通道突然变窄的缝隙处（图 6.3），叶轮带动水流旋转，当叶轮外圆周速度较大时，间隙内水流迅速降压并产生汽化，形成汽蚀。间隙汽蚀对水泵运行的影响是噪声、长期击打。只要间隙在许可范围内，对水泵运行性能影响不大。如轴流泵叶片外缘与泵壳之间、离心泵减漏环与叶轮外缘之间均易发生间隙汽蚀。

3. 漩涡型汽蚀

当前池、进水池及进水流道中出现不良流态时，容易诱发漩涡，当漩涡中心的压力降低到汽化压力以下时，就会产生漩涡型汽蚀（图 6.4）。通常漩涡运动是不稳定的，所以这种汽蚀也是不稳定的、瞬息多变的。当漩涡接触到叶片或壁面时，才对材料产生汽蚀破坏，而这种汽蚀对流场的压力和流速分布影响很大，因而将严重危及水泵的稳定运行。

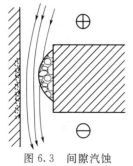

图 6.3　间隙汽蚀

Fig. 6.3　Clearance cavitation

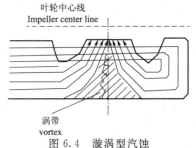

叶轮中心线
Impeller center line

涡带
vortex

图 6.4　漩涡型汽蚀

Fig. 6.4　Vortex cavitation

6.4　水泵中汽蚀的危害

1. 水泵工作性能恶化

泵内产生汽蚀时，大量气泡的存在，改变了流道内特别是叶槽内过流面积和流动方向，而使叶轮与水流之间能量交换的稳定性遭到破坏，能量损失增加，引起流量、扬程及效率的下降，甚至达到断流状态。对于不同比转速的水泵，汽蚀对工作性能产生的影响不同。低比转速离心泵由于叶槽狭长，宽度较小，一旦汽蚀发生，气泡即迅速扩展到整个叶槽，引起断流，泵的特性曲线呈急剧下降形状，如图 6.5（a）所示。对于混流泵，由于

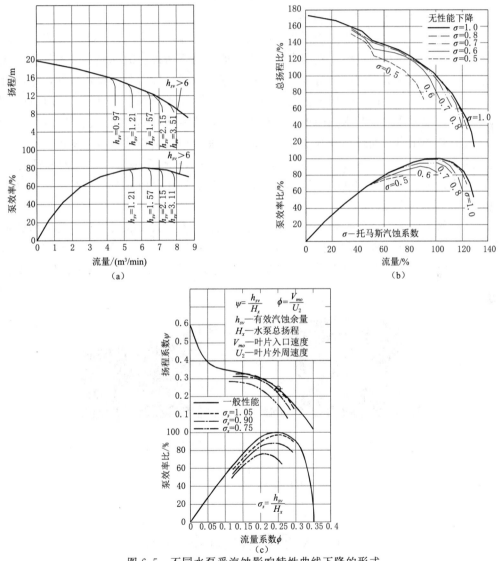

图 6.5　不同水泵受汽蚀影响特性曲线下降的形式

Fig. 6.5　Form of drop in characteristic curve of different pumps affected by cavitation

（a）离心泵；（b）混流泵；（c）轴流泵

(a) Centrifugal pump；(b) Mixed flow pump；(c) Axial flow pump

叶轮槽道较宽，因此当汽蚀发生时，只是发生在叶槽的某一局部，随着汽蚀状态的发展，才会布满整个叶槽，在特性曲线上则表现为开始时下降较和缓，最后呈迅速下降之势，如图 6.5（b）所示。至于高比转速的轴流泵，由于叶片之间流道相当宽阔，故汽蚀区不易扩展到整个叶槽，特性曲线下降缓慢，以至无明显的断裂点，如图 6.5（c）所示。

2．发生噪声和振动

水泵发生汽蚀时，气泡连续产生、膨胀和溃灭而产生脉冲压力，发出类似炒黄豆的噼噼啪啪的声音，严重时水泵周围发出难以忍受的噪声。此外，还伴随产生频率很高的振动，泵尺寸越大，噪声和振动越严重，强烈的振动会使水泵部件和机座受到破坏而缩短寿命。

3．水泵过流表面材料产生汽蚀破坏

当水泵发生空化后，只有对固体边壁的损伤和破坏才能称为汽蚀。接近和附着在边壁的气泡破裂后形成很小的空隙，周围水体在较高的压力下通过空隙形成了微型射流，高速冲击边壁。根据高速摄影实验，射流速度每秒可达几百米，冲击时间仅几微秒，作用面积又非常小，因此对边壁产生的压力很大。长期冲击则造成金属疲劳损失。还有观点认为，气泡溃灭时产生强烈的瞬时水锤冲击波。在无数气泡溃灭下，冲击频率达到每秒几万次。这一巨大的脉冲压力长期作用在水泵部件壁面上，使金属产生塑性变形和硬化，然后在冲击压力的持续作用下，表面金属颗粒逐步剥落，最后形成蜂窝麻点状的孔洞和裂缝（图6.6），直至壁面整个破坏和断裂。汽蚀发生时还伴随着氧化反应和电化学作用，而以机械剥蚀为主，三者相互促进。

金属材料抗汽蚀性能的好坏对汽蚀破坏速率影响很大，我国农用水泵叶轮大部分是铸铁和碳钢制成，表面粗糙度差，抗汽蚀性能很差，往往运行几百小时就被剥蚀破坏。随着逐步采用高强度不锈钢材料和精密数控加工工艺，水利用大中型水泵的抗汽蚀性能明显改善。

（a）　　　　　　　　　　　（b）

图 6.6　叶片进口、叶轮室中部的汽蚀破坏

Fig. 6.6　Cavitation damage at blade inlet and middle of casing

（a）叶片进口；（b）叶轮室间隙

(a) Blade inlet；(b) Casing clearance

6.5　水泵汽蚀基本方程

泵在运行时是否发生汽蚀，是由外界调节、吸入装置及泵自身特性等因素共同决定的。现从水流能量观点出发，推导水泵汽蚀基本方程，从而得到水泵内发生汽蚀的物理条

件以及影响汽蚀的各个因素之间的关系，以提供采用抗汽蚀措施的理论依据。

如图 6.7 所示水泵装置，从水源吸水，设吸水进水池水面上的压力为 p_a。选取四个典型断面：进水池池面、泵进口断面 $S—S$、泵叶轮进口前断面 $O—O$ 和泵叶轮内压力最低点所在的断面 $K—K$。以进水池液面为参考基准面，分别列进水池液面和断面 $O—O$ 的能量方程，以及断面 $O—O$ 和断面 $K—K$ 的相对运动能量方程即伯努利方程，有

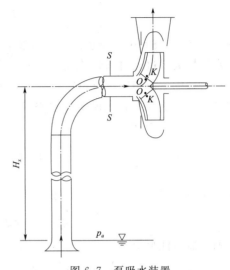

图 6.7 泵吸水装置

Fig. 6.7 Suction device of pump

$$\frac{p_a}{\rho g}=H_x+\frac{p_O}{\rho g}+\frac{v_O^2}{2g}+h_x+h_{s-O} \qquad (6.3)$$

$$z_O+\frac{p_O}{\rho g}+\frac{w_O^2}{2g}-\frac{u_O^2}{2g}-h_{O-K}=z_K+\frac{p_K}{\rho g}+\frac{w_K^2}{2g}-\frac{u_K^2}{2g} \qquad (6.4)$$

式中：p_O、p_K、v_O、u_O、w_O、u_K、w_K 为相应于断面 $O—O$、$K—K$ 的压力和速度；H_x 为进水池液面至泵进口断面 $S—S$ 的几何高度；h_x 为进水池液面至泵进口断面的水头损失；h_{s-O} 为断面 $S—S$ 至断面 $O—O$ 的水头损失；z_O、z_K 为相应于断面 $O—O$、$K—K$ 的位置高度；$h_{s-O}=\zeta_v\frac{v_O^2}{2g}$。

因 O，K 两点靠近，认为 $u_O=u_K$，$z_O=z_K$，$h_{O-K}=0$，不计水流的预旋，进口速度三角形为直角三角形，$w^2=v^2+u^2$，则式（6.4）变为

$$\frac{p_O}{\rho g}=\frac{p_K}{\rho g}+\frac{w_K^2-w_O^2}{2g} \qquad (6.5)$$

把式（6.5）代入式（6.3），得

$$\frac{p_a}{\rho g}-\frac{p_K}{\rho g}=H_x+h_x+\mu\frac{v_O^2}{2g}+\lambda\frac{w_O^2}{2g} \qquad (6.6)$$

式中：μ 为动能系数，$\mu=1+\zeta_v$；λ 为汽蚀系数，$\lambda=\frac{w_K^2}{w_O^2}-1$。

式（6.6）的物理意义是大气压力在泵内最低压力以上的剩余能量用于以下四个方面：

(1) 提升 H_x 的吸水高度。

(2) 克服吸水管路损失 h_x。

(3) 转为动能 $\frac{v_O^2}{2g}$。

(4) 满足水流在进入叶片后的压力下降 $\lambda\frac{w_O^2}{2g}$。

对式（6.6）做进一步整理，得

$$\frac{p_a}{\rho g}-\frac{p_K}{\rho g}-(H_x+h_x)=\mu\frac{v_O^2}{2g}+\lambda\frac{w_O^2}{2g} \qquad (6.7)$$

当 $p_K = p_{cav}$，汽化（以下标 cav 表示）开始，为汽蚀临界状态，上式变为

$$\frac{p_a}{\rho g} - \frac{p_{cav}}{\rho g} - (H_x + h_x)_{cav} = \left(\mu \frac{v_O^2}{2g} + \lambda \frac{w_O^2}{2g}\right)_{cav} \tag{6.8}$$

在式（6.8）中，左边的项仅与装置有关，表示水泵吸水侧装置能量扣除吸上高度和吸水管损失超过当时当地汽化压力所剩的富余能量，称为有效汽蚀余量，用 $NPSH_a$ 表示；右边的项仅与叶轮进口结构有关，表示叶轮进口附近的动压降，称为必需汽蚀余量，用 $NPSH_r$ 表示。该式说明在水泵装置的有效汽蚀余量与水泵必需汽蚀余量相等时，水泵进入汽蚀临界状态。

在定义上
$$NPSH_r = \left(\mu \frac{v_O^2}{2g} + \lambda \frac{w_O^2}{2g}\right)_{cav} \tag{6.9}$$

在数值上
$$NPSH_r = \frac{p_a - p_{cav}}{\rho g} - (H_x + h_x)_{cav} \tag{6.10}$$

式（6.8）～式（6.10）即为汽蚀基本方程。在汽蚀试验中常用式（6.10）来计算水泵的必需汽蚀余量。

叶轮进口前后压差（压降）取决于泵的运行工况和泵进口结构。一般而言，泵内动压降 $\lambda \frac{w_O^2}{2g}$ 数值较大。因相对速度 w 由绝对速度 v 和圆周速度 u 决定，故对于低比转速水泵，动压降数值较小，此时在汽蚀余量中动能项 $\frac{v_O^2}{2g}$ 占主导地位；而对于高比转速水泵，动压降 $\lambda \frac{w_O^2}{2g}$ 较大，是汽蚀的主要因素。

6.6　汽蚀余量的计算

汽蚀余量是反映水泵汽蚀性能的重要参数，并用于计算水泵吸水高度和安装高程，在实践中具有重要意义。汽蚀余量包括泵必需汽蚀余量和装置有效汽蚀余量两种。

6.6.1　装置有效汽蚀余量计算

装置有效汽蚀余量是指水泵进口处液体所具有的超过当时温度下汽化压力的富裕能量，用 $NPSH_a$ 表示，用公式表达为

$$NPSH_a = \frac{P_S}{\rho g} + \frac{v_S^2}{2g} - \frac{P_{cav}}{\rho g} \tag{6.11}$$

式中：$\frac{P_S}{\rho g}$ 为水泵进口 S—S 断面具有的压力水头，m；$\frac{v_S^2}{2g}$ 为水泵进口 S—S 断面具有的流速水头，m；$\frac{P_{cav}}{\rho g}$ 为汽化压力值，见表 6.1。

由图 6.7，列出从进水面和泵进口 S—S 断面的能量方程

$$\frac{P_a}{\rho g} - H_x - h_x = \frac{P_S}{\rho g} + \frac{v_S^2}{2g} \tag{6.12}$$

式中：$\frac{P_a}{\rho g}$ 为大气压力水柱高，从表 6.2 中选取；H_x 为水泵的实际吸水高度（安装高度），

即进水池液面到水泵基准面的垂直距离，m；h_x 为从进水面至泵进口的水力损失，m。

表 6.2　　　　　　　　　　　　　　　　不同海拔时的大气压力

Table 6.2　　　　　　　　　　　　**Atmospheric pressure at different altitudes**

海拔 altitude/m	−600	0	100	200	300	400	500	600
大气压力 atmospheric pressure $(Pa/\rho g)$ /m	11.3	10.33	10.22	10.11	9.97	9.89	9.77	9.66
海拔 altitude/m	700	800	900	1000	2000	3000	4000	5000
大气压力 atmospheric pressure $(Pa/\rho g)$ /m	9.55	9.44	9.33	9.22	8.11	7.47	6.52	5.57

各种不同安装方式的水泵基准面的选取如图 6.8 所示。

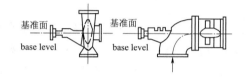

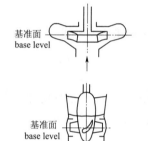

图 6.8　水泵的安装基准面

Fig. 6.8　Installation base level of the pump

将式（6.12）代入式（6.11），有

$$NPSH_a = \frac{P_a}{\rho g} - H_x - h_x - \frac{P_{cav}}{\rho g} \tag{6.13}$$

由式（6.13）可见：装置有效汽蚀余量的大小只与水泵装置进水侧的情况有关，即只与作用于液面的大气压力、水泵安装高度、水泵装置进水部分的水力损失及当时温度的汽化压力水头等因素有关，而与水泵本身的因素无关。

6.6.2　泵必需汽蚀余量的计算

泵必需汽蚀余量是水泵汽蚀性能的基本工作参数，用 $NPSH_r$ 表示。由式（6.9）可知，水泵的必需汽蚀余量为泵进口处水流在被叶轮加压前所产生的不可避免的动压降。该式将入口速度参数与泵的汽蚀性能直接联系起来，理论上是严密的。但公式中的系数为经验系数，积累的数值较少，在非设计工况下变动较大，因此目前一般采用试验法、托马斯汽蚀系数法、汽蚀比转速法求得 $NPSH_r$。

水泵厂向用户提供的水泵汽蚀余量，一般是偏安全的，即用该泵发生汽蚀时的汽蚀余量加一个安全量确定的，称为该泵的允许必需汽蚀余量，用 $[NPSH_r]$ 表示。

6.6.3　泵汽蚀发生与汽蚀余量

水泵是否发生汽蚀，取决于泵装置的条件与泵自身条件。

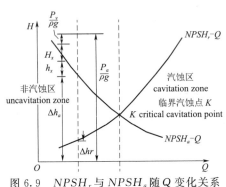

图 6.9　$NPSH_r$ 与 $NPSH_a$ 随 Q 变化关系

Fig. 6.9　Relationship between $NPSH_r$ and $NPSH_a$ with Q

由式（6.9）和式（6.10）可以看出，随着流量的增加，泵必需汽蚀余量增加，而装置有效汽蚀余量则减少，如图 6.9 所示。在两曲线的交点 K 的左边，表示装置有效汽蚀余量大于泵必需汽蚀余量，水泵不会发生汽蚀；K 点的右边，表示装置有效汽蚀余量小于泵必需汽蚀余量，水泵将会发生汽蚀。K 点称为临界汽蚀点。

根据 $NPSH_a$ 与 $NPSH_r$ 可判断水泵是否发生汽蚀：当 $NPSH_a > NPSH_r$ 时，水泵不发生汽蚀；当 $NPSH_a < NPSH_r$ 时，水泵发生汽蚀。

6.7　汽蚀相似律与汽蚀比转速

6.7.1　汽蚀相似律

汽蚀相似律是反映一系列几何相似的水泵在相似工况时汽蚀性能之间的关系。两台相似的水泵，根据水泵的相似理论可推得它们的汽蚀余量之间有如下的关系

$$\frac{NPSH_{rP}}{NPSH_{rM}} = \left(\frac{n_P D_P}{n_M D_M}\right)^2 \tag{6.14}$$

式中：下标 P 和 M 分别表示原型泵和模型泵；n 和 D 则分别表示水泵的转速和线性尺寸（一般取叶轮直径）。

式（6.14）称为水泵的汽蚀相似律。

式（6.14）表明：相似水泵的必需汽蚀余量之比与它们的线性尺寸之比的平方成正比，与转速比的平方成正比。可见汽蚀余量的相似性类似于扬程的相似性。在相似水泵中，转速、尺寸越大，$NPSH_r$ 也越大，抗汽蚀能力越差。对于同一台水泵，当水泵的转速改变时，必需汽蚀余量也随之改变，由于叶轮直径不变，由式（6.14）得

$$\frac{NPSH_{r1}}{NPSH_{r2}} = \left(\frac{n_1}{n_2}\right)^2 \tag{6.15}$$

式中：下标 1、2 分别表示工况 1 和工况 2。

6.7.2　汽蚀比转速

汽蚀比转速是表明水泵汽蚀性能的重要参数，也是汽蚀相似性的重要判据，可由水泵的相似理论推求出水泵的汽蚀比转速 C。

$$C = \frac{n\sqrt{Q}}{NPSH_r^{3/4}} \qquad (6.16)$$

若 n 的单位取 r/min，Q 的单位取 m³/s，$NPSH_r$ 的单位取 m，则上式需要乘以系数 5.62，即

$$C = 5.62\frac{n\sqrt{Q}}{NPSH_r^{3/4}} \qquad (6.17)$$

从上式可知，C 值越大，$NPSH_r$ 越小，水泵汽蚀性能越好。所以，由 C 值的大小可以很容易判别出水泵汽蚀性能的优劣。与比转速一样，式中双吸泵汽蚀比转速计算流量取泵流量的二分之一。

水泵的汽蚀比转速和比转速均为水泵的相似准则。水泵的比转速 n_s 是水泵的出口相似准则，用于水泵的能量性能换算；而水泵的汽蚀比转速 C 是水泵的进口相似准则，用于水泵的汽蚀性能换算。

6.8 水 泵 安 装 高 度

水泵的安装高度 H_x 指水泵的基准面（图 6.8）到进水池水面的垂直高度，是设计泵站各个高程的基本依据。H_x 可由泵的允许吸上真空度 H_s 或必需汽蚀余量 $NPSH_r$ 来计算，而 H_s 和 $NPSH_r$ 均由泵的汽蚀性能来决定，在泵样本上通常标注该值。

6.8.1 允许吸上真空度

水泵的吸上真空度是指水泵进口处的压强低于大气压强的数值，如果进水池液面的压强是一个大气压，则水泵进口处的真空度就是大气压强与该泵进口 S—S 断面压强的差值，通常在 S—S 断面处安装真空表，真空表上的读数就是水泵的真空度，用米水柱高作单位，则有

$$H_s = \frac{P_a}{\rho g} - \frac{P_s}{\rho g} \qquad (6.18)$$

式中：$\frac{P_s}{\rho g}$ 为泵进口 S—S 断面的压力，m。

水泵吸上真空高度也是反映水泵汽蚀性能的参数，吸上真空高度越大，说明该泵的抗汽蚀能力越好。

对于离心泵，一般用吸上真空高度值作为反映泵的汽蚀性能的参数，其值由试验确定。为了安全起见，往往在出厂时把它的许用值定得小一点，即将该泵刚刚发生汽蚀时的吸上真空高度值减去一个安全量（0.3～0.5m），称为泵的允许吸上真空高度，用 $[H_s]$ 表示。

6.8.2 离心泵、蜗壳式混流泵安装高度计算

1. 用允许吸上真空高度 $[H_s]$ 计算

卧式离心泵和蜗壳式混流泵一般安装在水面以上。由汽蚀临界状态时的 $[H_s]$：

$$[H_s] = H_x + h_x + \frac{v_s^2}{2g}$$

85

则有

$$H_x = [H_s] - h_x - \frac{v_s^2}{2g} \tag{6.19}$$

2. 用允许必需汽蚀余量 $[NPSH_r]$ 计算

如提供的泵汽蚀参数是 $[NPSH_r]$，根据临界状态 $[NPSH_r]$：

$$[NPSH_r] = \frac{P_a}{\rho g} - H_x - h_x - \frac{P_{cav}}{\rho g}$$

则有

$$H_x = \frac{P_a}{\rho g} - h_x - [NPSH_r] - \frac{P_{cav}}{\rho g} \tag{6.20}$$

常温下简化时采用

$$H_x = 10.0 - h_x - [NPSH_r] \tag{6.21}$$

因为在 $[NPSH_r]$ 中已考虑了安全量，计算出来的 H_x 值可不必再考虑取安全余量。

3. 海拔及汽化压力的修正

必须注意的是：泵的允许吸上真空高度指海拔为 0 和环境温度为 20℃ 的标准状况时的值。如果使用地点和温度均为非标准状况，则允许吸上真空度值要按下式修正

$$[H_s]' = [H_s] + \frac{P_a}{\rho g} - 10.33 - \frac{P_{cav}}{\rho g} + 0.24 \tag{6.22}$$

式中：$[H_s]'$ 为海拔、温度修正之后的允许吸上真空高度，m；$[H_s]$ 为标准状况时的允许吸上真空高度，m；$\frac{P_a}{\rho g}$ 为水泵安装地点的大气压力，m，见表 6.2；$\frac{P_{cav}}{\rho g}$ 为工作温度下的汽化压力，m，见表 6.1。

上式中算出的 H_x 若大于 0，表示该泵的基准面在水面以上，说明该泵有正吸程。若 H_x 小于 0，则说明该泵的基准面必须在水面以下，也就是说泵的叶轮必须淹没在水下，水泵吸程为负值。

6.8.3　轴流泵、导叶式混流泵安装高度计算

因轴流泵、导叶式混流泵进水管路较短，中小型泵站只有一节进水喇叭，故进水损失可以忽略不计，由式（6.21）得

$$H_x = 10.0 - [NPSH_r] \tag{6.23}$$

为便于启动，即使计算得出的 H_x 值为正值，也不将叶轮安装在水面以上，其叶轮基准面离水面的深度，一般为 1.2 倍的进水喇叭直径或不小于 0.5~1m。

6.9　泵汽蚀发生原因与防护

泵站在运行中，一定程度的汽蚀现象是很难避免的，其主要原因如下：

（1）泵在非设计工况下运行概率较大。

（2）即便泵站吸入装置设计合理，但某些部位的汽蚀仍然可能产生，如叶片与泵壳处、口环与蜗壳处的间隙汽蚀。

（3）多泥沙水源对水泵的磨蚀作用加剧了空化现象和汽蚀破坏。

（4）泵的高速化、大流量带来了汽蚀的严重性。

（5）泵站水泵选型不当，导致水泵长期偏离设计工况运行。

防止水泵汽蚀，可以从改进水泵和使用条件两方面着手。

（1）提高水泵抗汽蚀能力，减少泵必需汽蚀余量。

1）改善泵的设计和制造工艺。

2）对已经被汽蚀损坏的过流部件，可采用表面保护技术加以修复。即采用具有一定硬度和韧性的材料，或对质地较差的部件进行表面强化处理，使其表面的硬度和韧性增加。表面保护技术可以显著提高水泵叶片、叶轮室的抗汽蚀、磨蚀的能力，并可延长其使用寿命。

3）对较重要的水泵，采用不锈钢等抗汽蚀材料制作水泵。

（2）改善水泵吸入装置和运行条件，增加装置有效汽蚀余量。

1）在布置水泵进水管路时，要尽量减少不必要的管路附件，例如进口的底阀、弯头、闸阀等，管路不要过长，进口管径可选大一些，降低管中流速，减小水头损失。

2）合理设计泵站进水池及引水建筑物，使水流平顺地进入水泵，不要发生偏流、回流、漩涡等不良流态，避免水泵进水水流挟气。必要时还要进行模型试验，采取必要的防涡措施，以改善进口水流状态。

（3）运行中要及时清理拦污栅，避免在泵站进水中产生过大的水位落差。

（4）必要时，在满足运行要求的前提下，可适当降低水泵转速，因为泵的汽蚀余量与转速比的平方成正比。

（5）当发生汽蚀时，可在吸入侧采用应急补入少量空气措施，以减小空化发生区的真空度，但必须严格控制补气量和补气时长。

Chapter 6   Vaporization and Cavitation in Pumps

Cavitation are abnormal phenomena in hydraulic machinery, hydraulic structures, ship propellers and other water-passing components. In 1874, Reynolds first theoretically described and observed the cavitation phenomenon in the laboratory, capturing the movement law of vapor-bubble in warm water. In 1895, when engineers inspected the propeller of an oil tanker crossing the Atlantic Ocean, they firstly found that the surface of propeller of an ocean ship was covered with small holes, which were seriously eroded into spongy shape, or even perforated, and even caused the whole blade to fall of and peeling off in more serious cases. In 1897, Barnaby and Parsons proposed the concept of "cavitation" after the occurrence of a severe decrease in propeller efficiency in succession of the "Courage" torpedo boat and several steam engine ships, and pointed out that cavitation may occur in the presence of high-speed relative motion between liquids and objects. Since then, people have realized the cavitation of water and the actual cavitation. Later, this phenomenon was found not only in propellers, but also in axial flow pumps, centrifugal pumps, turbines, deep-water weapons and even pipes.

Vaporization and cavitation occur during the operation of the pump, which will deteriorate the performance of the pump until the discharge is cut off, or cause honeycomb pitting of the impeller and the pump easing and other water-passing components to be destroyed. The cavitation performance of the pump is also a decisive factor in determining the installation elevation of the pump. Therefore, great attention must be paid to the problems of vaporization and cavitation in pumps.

6. 1   Cavitation mechanism

6. 1. 1   Vaporization and cavitation of water

At a certain pressure, when the temperature rises to a certain critical value, vaporization occurs and bubbles appear, i. e. vapor phase occurs in the water body, called boiling, this critical temperature is called boiling point. Other liquids produce the same phenomenon, except that the boiling points are different.

In contrast to boiling, for a certain liquid, when its temperature is fixed and the pressure in the liquid decreases to a certain critical value, vaporization will also occur in the liquid, generating numerous bubbles filled with steam and gas, and the continuity of the liquid is destroyed. This process of forming the gas phase due to the reduction of pressure is called cavitation phenomenon. The critical pressure corresponding to vaporization is called saturated vapor pressure or vaporization pressure. Different water temperatures have different vaporization pressures. Table 6. 1 lists the vaporization pressure values at different water temperatures, expressed as water column height in meters.

The two phenomena of boiling and cavitation are identical in physical essence, at least in their initial problems, but the reasons for the formation of vaporization of liquids are different. Boiling is caused by heating up, while cavitation is caused by pressure decreases. Cavitation can generally be attributed to

$$P \leqslant P_{cav} \tag{6.1}$$

The above explanation of cavitation due to reduced pressure in liquids is often adopted by the engineering community. In hydraulic machinery, the reduction of water pressure is often caused by excessive flow velocity, so a dimensionless number combined by pressure and flow velocity is often used to express the degree of potential cavitation of flow, which is called the number of cavitation of flow, and its expression is as follows

$$\sigma = \frac{p_\infty - p_v}{\frac{1}{2}\rho v_\infty^2} \tag{6.2}$$

Where, p_∞ and v_∞ are the pressure and velocity at the undisturbed point in the flow stream respectively; p_v is the pressure at some point in the flow stream.

Cavitation occurs when the pressure p_v decreases to the critical pressure P_{cav}. The value σ at initial cavitation can be called the incipient cavitation number σ_i. Generally speaking, the larger the number of cavitation, the less prone to cavitation. For pump blades, $\sigma_i = 0. 1 - 0. 2$.

6. 1. 2 Cavitation phenomenon

If the absolute pressure of overflow area decreases to the vaporization pressure at the temperature in local area, the water will vaporize here, producing steam and forming bubbles. These bubbles flow forward with the water flow, and when they reach the high pressure, the high pressure liquid around the bubbles causes the bubbles to shrink sharply to burst (condensation) . While the bubbles condense, the liquid particles will impact at high speed. At the center of the hole, these particles collide with each other to produce local high pressure, accompanied by the sound of bursting. If the bubble bursts just at the solid surface, this kind of impact will directly act on the solid surface. Because the contact area is very small, the pressure value is extremely high, and the frequency of the bubble collapse is extremely high, which can reach tens of thousands of times per second, and the impact on the solid surface is extremely great, resulting in fatigue peeling of the solid surface, honeycomb-shaped pits appear, and then peeling off, even breaking down in severe cases. This phenomenon is called "cavitation" .

Vaporization and cavitation are very complex physical and chemical phenomena. Although the two phenomena are closely related, their macroscopic characteristics and mechanisms are still different. The main differences are as follows.

(1) Vaporization is a fluid dynamic phenomenon, which is the process of the development, expansion and collapse of countless microbubbles, while cavitation is the mechanical damage when the bubbles collapse, chemical corrosion and electrochemical effects on the side wall materials and so on.

(2) Vaporization occurs in the low pressure area (the space less than the vaporization pressure), and cavitation occurs in the high pressure area (the pressure rise space).

(3) The starting point of cavitation is usually after the starting point of cavitation, sometimes far away.

(4) From the above, it can be further deduced that for specific boundaries, cavitation does not necessarily have cavitation. For example, for the impeller of a super cavitation pump, cavitation can be said to be avoided to a certain extent.

(5) In the same vaporization state, the solid boundary conditions and material properties are different, and the strength of cavitation may be different.

6. 2   External hydraulic factors inducing pump cavitation

The liquid pressure in the local area when the pump is running is lower than the vaporization pressure under the operating temperature, which is the condition that cavitation occurs and then causes cavitation. Besides the structure and performance factors of the pump itself, the external hydraulic factors are as follows.

1. Low pressure of suction system

(1) The excessive suction height of the pump increases the vacuum at the inlet of the pump and decreases the absolute pressure.

(2) The high altitude and low atmospheric pressure at the place where the pump is installed cause the pressure of the suction unit to drop.

(3) The pressure of suction unit decreases with the increase of pump discharge and flow velocity.

(4) The liquid being sucked has higher temperature and higher vaporization pressure, so it is easy to vaporize.

2. Low pressure in local area of pump

(1) When the operation of the pump deviates from the design conditions, impact and swirl occur on the water flow entering the impeller, which increase the local discharge and reduce the local pressure.

When the pump is running at a working condition over the design discharge, the relative velocity direction of inlet deviates from the rotating direction of impeller and β_1 increases, whirlpool occurs on the front of blade leading edge, which is prone to negative pressure and vaporization. When the pump operates under the condition of less than the de-signed flow, the relative velocity direction of inlet deviates from the reverse direction, β_1 decreases, and the whirlpool occurs on the back of the blade, which aggravates the vapori-zation degree of the low pressure area on the back of the blade (see Fig. 6.1).

(2) In the axial flow pump with elbow or oblique inlet passage, the inner side of the flow curve is easy to flow off, resulting in uneven distribution of the inlet flow velocity of the impeller of the pump (see Fig. 6.2). When the blade rotates to the outside of the curve, a negative pressure zone is formed on the front of the leading edge of the blade. When the blade rotates to the inside, a negative pressure zone is formed on the back of the blade. Therefore, if the inlet flow pattern is not good, the velocity of flow distribution at the outlet of the inlet flow channel is uneven, it is possible to cause the negative pressure on both the front and the back of the blade to cause cavitation.

(3) When the water flow in the pump passes through the narrow gap, such as the gap between the outer edge of the impeller of the axial flow pump and the pump casing, the separator tongue of the volute of the centrifugal pump, etc., the pressure at the gap decreases due to the large discharge, thus reaching the vaporization condition.

(4) The casting quality of the surface of the over-flow part of the pump is poor, and local eddies are generated behind some rough protrusions, causing local pressure drop.

(5) Vibration of some parts in water occurs at a certain frequency due to the vibration of the unit. Positive and negative pressures alternate on both sides of the parts. Vibratory cavitations occur when the negative pressure reaches the vaporization pressure.

6. 3 Cavitation types

In the pump cavitation test, the cavitation in the pump impeller can be clearly ob-served through the observation window with a strobometer, and the cavitation process can also be recorded with a high-speed camera. The experiments show that the cavitation can be divided into migration type, fixed type, vortex type according to the cavitation form. According to the location of cavitation in pump, it can be roughly divided into three types: airfoil cavitation, clearance cavitation and vortex cavitation in flow channel.

1. Airfoil cavitation

Cavitation on the front and reverse side of a blade due to its airfoil type is called airfoil cavitation. Cavitation is caused by the deviation of relative speed direction of blade inlet from the design direction when the pump operates under non-design conditions, resulting in the reduction of pressure on the front or reverse side of the blade.

With the change of pump operating conditions, the types of airfoil cavitation change, which can be divided into the following three types:

(1) Class I cavitation occurs at the back of the blade inlet at the outer edge of the

impeller, when the discharge is less than the discharge at the highest efficiency point, when the discharge continues to decrease, the head continues to increase, the blade inlet angle increases, and the cavitation area on the back of the blade gradually expands to the outlet edge and the blade root.

(2) Class II cavitation occurs at the outer edge, the root and the front of the inlet of the impeller. At this time, the discharge is greater than the discharge at the BEP. When the flow continues to increase, the head continues to decrease, the blade inlet angle decreases, and the cavitation area on the front of the blade extends to the middle.

(3) Class III cavitation occurs at the outer edge and root of the blade, reaching most areas of the blade from the near front end of the blade to the rear. At this time, the blade angle is large and the discharge is large. Due to the large area where cavitation occurs, the performance of pumps is often degraded, which is called "harmful cavitation".

2. Clearance cavitation

Clearance cavitation mainly occurs when water flow passes through a sudden narrowing gap (see Fig. 6.3). The impeller drives the water flow to rotate. When the outer circumferential velocity of the impeller is large, the water flow in the gap rapidly reduces pressure and produces vaporization, forming cavitation. The effect of clearance cavitation on pump operation is noise and strike. As long as the clearance is within the allowable range, it has little effect on pump performance. For example, clearance cavitation is easy to occur between the outer edge of the blade of the axial flow pump and the pump shell, and between the leakage reducing ring of the centrifugal pump and the outer edge of the impeller.

3. Vortex cavitation

When there are bad flow patterns in the forebay, sump and inlet passage, it is easy to induce vortexes. When the pressure in the center of the vortexes decreases below the vaporization pressure, vortex cavitation will occur (see Fig. 6.4). Usually the vortex motion is unstable, so the cavitation is also unstable and transient. When the vortex touches the blade or wall, it will cause cavitation damage to the material, and this kind of cavitation has a great impact on the pressure and velocity distribution of the flow field, which will seriously endanger the stable operation of the pump.

6.4   Hazards of cavitation

1. Worsening of pump performance

When cavitation occurs in the pump, due to the existence of a large number of bubbles, the area and direction of flow in the flow channel, especially in the vane groove, are changed, and the stability of energy exchange between impeller and water flow is damaged, the energy loss is increased, resulting in the decrease of discharge, head and efficiency, and even to the state of cut-off. For pumps with different specific speeds, the effect of cavitation on working performance is different. For low specific speed centrifugal pump, due to the narrow and long blade groove and small width, once cavitation occurs, the bubble expands rapidly to the whole blade groove, causing a cut-off, and the characteristic curve of the pump shows a sharp drop shape, as shown in Fig. 6. 5 (a). For mixed flow pump, the impeller channel is wider, when cavitation occurs, it only occurs in a part of the blade groove. With the development of cavitation, the whole blade groove will be filled. On the characteristic curve, it shows that the decline is gentle at the beginning and rapid at the end, as shown in Fig. 6. 5 (b). As for the axial flow pump with high specific speed, because the flow path between the blades is quite wide, the cavitation zone is not easy to extend to the entire groove, and the characteristic curve drops slowly, so that there is no obvious breakpoint, as shown in Fig. 6. 5 (c).

2. Noise and vibration

Pump cavitation, due to the continuous generation, expansion and collapse of air bubbles to generate pulse pressure. It sounds like fry beans, seriously, the surrounding of the pump emits unbearable noise. In addition, accompanied by high frequency vibration, the larger the size of the pump, the more serious the noise and vibration, strong vibration will damage the pump components and seat and shorten the life.

3. Erosion of flow channel material

When the pump vaporization occurs, only the damage and destruction of the solid side wall can be called cavitation. When the bubbles close to and attached to the side wall burst, a small gap is formed, and the surrounding water body passes through the gap under a high pressure to form a micro jet, which impinges on the side wall at a high speed. According to high-speed photography experiments, the jet speed can reach several hundred meters per second, the impact time is only a few microseconds, and the working area is very small, so the pressure on the side wall is very high. Long-term impact results in metal fatigue loss. It is also conceived that a strong instantaneous water hammer shock wave occurs when a bubble collapses. With countless bubble bursts, the shock frequency is significantly tens of thousands of times per second. This huge impulse pressure acts on the wall of the pump components for a long time, causing plastic deformation and hardening of the metal. Then, under the continuous impact pressure, the surface metal particles gradually flake off, finally forming honeycomb pitted holes and cracks (Fig. 6. 6), until the whole wall is destroyed and broken. Cavitation is accompanied by oxidation and electrochemical reaction, while mechanical erosion is the main one, which promotes each other.

The anti-cavitation performance of metal material has a great influence on the rate of cavitation damage. Most of the impellers of agricultural pumps in China are made of cast iron and carbon steel. Their surface roughness is poor and their anti-cavitation performance is poor, and they are often destroyed after running for several hundred hours.

With the gradual adoption of high-strength stainless steel material and precise CNC (computer numerical control) machining process, the cavitation resistance of large and medium-sized pumps has been significantly improved.

6. 5　Basic equation of pump cavitation

Whether cavitation occurs during operation of pump is determined by external adjustment, suction system and pump characteristics. From the point of view of water flow energy, the basic equation of pump cavitation is deduced, to get the physical conditions of cavitation in the pump and the relationship between various factors affecting cavitation, to provide theoretical basis for anti-cavitation measures.

When water is absorbed from the water source, a pressure of p_a on the water surface of the suction sump is set as shown in Fig. 6. 7. Four typical sections are selected: sump surface, pump inlet section $S—S$, pump impeller inlet front section $O—O$ and pump impeller inner pressure at the lowest point of $K—K$ section. With the intake pool level as the reference datum, the energy equations of the intake pool level and the $O—O$ section, as well as the relative motion energy equations of the $O—O$ section and the $K—K$ section, namely the Bernoulli equation, are listed respectively.

$$\frac{p_a}{\rho g} = H_x + \frac{p_O}{\rho g} + \frac{v_O^2}{2g} + h_x + h_{S-O} \tag{6.3}$$

$$z_O + \frac{p_O}{\rho g} + \frac{w_O^2}{2g} - \frac{u_O^2}{2g} - h_{O-K} = z_K + \frac{p_K}{\rho g} + \frac{w_K^2}{2g} - \frac{u_K^2}{2g} \tag{6.4}$$

Where, p_O, p_K, v_O, u_O, w_O, u_K and w_K are the pressures and velocities corresponding to sections $O—O$ and $K—K$; H_x is the geometric height of the inlet pool water level to the pump inlet section $S—S$; h_x is the head loss from the inlet pool water level to the pump inlet section; h_{S-O} is the head loss from section $S—S$ to section $O—O$; z_O, z_K are corresponding to sections O、K position height; $h_{S-O} = \zeta_v \frac{v_O^2}{2g}$.

Owing to the close proximity of O and K, $u_O = u_K$, $z_O = z_K$, $h_{O-K} = 0$, the inlet velocity triangle is a right triangle regardless of the pre-rotation of water flow, $w^2 = v^2 + u^2$, then Eq. (6.4) becomes

$$\frac{p_O}{\rho g} = \frac{p_K}{\rho g} + \frac{w_K^2 - w_O^2}{2g} \tag{6.5}$$

Substitute Eq. (6.5) into Eq. (6.3) to obtain

$$\frac{p_a}{\rho g} - \frac{p_K}{\rho g} = H_x + h_x + \mu \frac{v_O^2}{2g} + \lambda \frac{w_O^2}{2g} \tag{6.6}$$

Where, μ is the kinetic energy coefficient, $\mu = 1 + \zeta_v$; λ is the cavitation coefficient, $\lambda = \frac{w_K^2}{w_O^2} - 1$.

The physical meaning of Eq. (6.6) is that the residual energy of atmospheric pressure above the minimum pressure in the pump is used in the following four aspects:

(1) Raise the water absorption height of H_x.

(2) Overcome the loss of h_x in suction pipeline.

(3) Convert to kinetic energy $\dfrac{v_O^2}{2g}$.

(4) Satisfy the pressure drop of water flow after it enters the blades, $\lambda \dfrac{w_O^2}{2g}$.

Eq. (6.6) is further sorted out to obtain

$$\frac{p_a}{\rho g} - \frac{p_K}{\rho g} - (H_x + h_x) = \mu \frac{v_O^2}{2g} + \lambda \frac{w_O^2}{2g} \tag{6.7}$$

When $p_K = p_{cav}$, vaporization (indicated by subscript cav) begins, which is the critical state of cavitation, the above formula becomes

$$\frac{p_a}{\rho g} - \frac{p_{cav}}{\rho g} - (H_x + h_x)_{cav} = \left(\mu \frac{v_O^2}{2g} + \lambda \frac{w_O^2}{2g} \right)_{cav} \tag{6.8}$$

In Eq. (6.8), the left item is only related to the device, which means that the energy of the device on the suction side of the pump deducts the suction height and the loss of the suction pipe exceeds the surplus energy left by the local vaporization pressure at that time. It is called available net positive suction head, which is expressed by $NPSH_a$. The right item is only related to the impeller inlet structure and represents the dynamic pressure drop near the impeller inlet, which is called required cavitation allowance and expressed by $NPSH_r$. This formula shows that when the $NPSH_a$ of the pump device is equal to the $NPSH_r$ of the pump, the pump enters the critical state of cavitation.

By definition
$$NPSH_r = \left(\mu \frac{v_O^2}{2g} + \lambda \frac{w_O^2}{2g} \right)_{cav} \tag{6.9}$$

Numerically
$$NPSH_r = \frac{p_a - p_{cav}}{\rho g} - (H_x + h_x)_{cav} \tag{6.10}$$

Eq. (6.8) – Eq. (6.10) are the basic equations of cavitation. The Eq. (6.10) is commonly used in the cavitation test to calculate the $NPSH_r$ of the pump.

The pressure difference (pressure drop) before and after the impeller inlet depends on the operation condition of the pump and the structure of the pump inlet. Generally speaking, the dynamic pressure drop in the pump is large. Because relative velocity w is determined by absolute velocity v and circumferential velocity u, the value of dynamic pressure drop $\lambda \dfrac{w_O^2}{2g}$ is small for low specific speed pump, at this time, the kinetic energy term $\dfrac{v_O^2}{2g}$ dominates in the $NPSH$; while for high specific speed pump, the dynamic pressure drop $\lambda \dfrac{w_O^2}{2g}$ is large, which is the main factor of cavitation.

6. 6   Calculation of *NPSH*

NPSH is an important parameter reflecting the cavitation performance of pump and is used to calculate the suction height and installation elevation of pump. It is of great significance in practice. *NPSH* includes $NPSH_r$ of pump and $NPSH_a$ of system.

6. 6. 1   *NPSH*$_a$

$NPSH_a$ refers to the abundant energy of the liquid at the inlet of the pump which exceeds the vaporization pressure at the temperature at that time. It is expressed by formula as follows

$$NPSH_a = \frac{P_S}{\rho g} + \frac{v_S^2}{2g} - \frac{P_{cav}}{\rho g} \tag{6.11}$$

Where, $\dfrac{P_S}{\rho g}$ is the pressure head of the $S-S$ section of the pump inlet, m; $\dfrac{v_S^2}{2g}$ is the velocity head of the $S-S$ section of the pump inlet, m; $\dfrac{P_{cav}}{\rho g}$ is the vaporization pressure value, see Table 6. 1.

From Fig. 6. 7, the energy equation of $S-S$ section from water surface of sump and pump inlet is listed

$$\frac{P_a}{\rho g} - H_x - h_x = \frac{P_S}{\rho g} + \frac{v_S^2}{2g} \tag{6.12}$$

Where, $\dfrac{P_a}{\rho g}$ is the height of atmospheric pressure water column, selected from Table 6. 2; H_x is the actual suction height (installation height) of the pump, i. e. the vertical distance from the sump level to the pump datum level, m; h_x is the hydraulic loss from the suction level to the pump inlet, m.

Selection of various pumps, installation methods and datum is shown in Fig. 6. 8.

The substitution of Eq. (6.12) for Eq. (6.11) is as follows

$$NPSH_a = \frac{P_a}{\rho g} - H_x - h_x - \frac{P_{cav}}{\rho g} \tag{6.13}$$

It can be seen from Eq. (6.13) that $NPSH_a$ is only related to the condition of the water inlet side of the pump system, It is only related to the atmospheric pressure acting on the liquid surface, the installation height of the pump, the hydraulic loss of the inlet part of the pump system and the vaporization pressure head of the temperature at that time, but have nothing to do with the factors of the pump itself.

6. 6. 2   *NPSH*$_r$

$NPSH_r$ is the basic working parameter of pump cavitation performance. By the Eq. (6.10), $NPSH_r$ of the pump is the inevitable dynamic pressure drop caused by the water flow at the pump inlet before being pressurized by the impeller. This formula directly links the inlet velocity parameters with the cavitation performance of the pump and is theoretically tight. However, the coefficients in the formula are empirical coefficients, which accumulate less values and vary greatly under non-design conditions. Therefore, at present, the $NPSH_r$ is generally obtained by the test method, the Thoma's cavitation coefficient method and the cavitation specific speed method.

$NPSH_r$ of pump provided by pump company to users is generally safe, that is, it is determined by adding a safety margin to the $NPSH_r$ of the pump, which is called the allowable $NPSH_r$ of the pump, and expressed by $[NPSH_r]$.

6.6.3　Pump cavitation and *NPSH*

Whether the pump has cavitation depends on the conditions of the pump system and the conditions of the pump itself.

It can be seen from Eq. (6.9) and Eq. (6.10) that with the increase of discharge, the $NPSH_r$ of the pump increases, while the $NPSH_a$ of the system decreases. The curve is shown in Fig. 6.9. On the left side of intersection point K of the two curves, it indicates that the $NPSH_a$ of the system is greater than the $NPSH_r$ of the pump, the pump will not cavitate, and on the right side of point K, it indicates that the $NPSH_a$ of the system is less than the $NPSH_r$ of the pump, and the pump will cavitate. K-point is called critical cavitation point.

According to $NPSH_a$ and $NPSH_r$, it can be judged whether the pump has cavitation: when $NPSH_a > NPSH_r$, the pump does not have cavitation; when $NPSH_a < NPSH_r$, the pump has cavitation.

6.7　Cavitation affinity law and cavitation specific speed

6.7.1　Cavitation affinity law

Cavitation affinity law reflects the relationship between a series of geometrically similar pumps and their cavitation performance under similar conditions. According to the affinity theory of two similar pumps, it can be inferred that there is the following relationship between their *NPSH*：

$$\frac{NPSH_{rP}}{NPSH_{rM}} = \left(\frac{n_P D_P}{n_M D_M}\right)^2 \tag{6.14}$$

Where subscripts P and M represent the prototype pump and the model pump, respectively; n and D represent the rotational speed and linear size of the pump respectively (generally taking impeller diameter).

Eq. (6.14) is called the cavitation affinity law of pump.

Eq. (6.14) shows that the ratio of the $NPSH_r$ of similar pumps is protortional to the square of the ratio of their linear dimensions, and proportional to the square of the speed ratio. It can be seen that the affinity of *NPSH* is similar to that of head. In similar pumps, the larger the speed and size, the larger the $NPSH_r$ and the worse the anti-cavitation ability. For the same pump, $NPSH_r$ also changes when the speed of the pump changes, because the impeller diameter is unchanged, obtained from Eq. (6.14)

$$\frac{NPSH_{r1}}{NPSH_{r2}}=\left(\frac{n_1}{n_2}\right)^2 \tag{6.15}$$

Where, subscripts 1 and 2 in the formula represent working conditions 1 and 2, respectively.

6.7.2 Specific speed of cavitation

Cavitation specific speed is an important parameter indicating the cavitation performance of pump, and also an important criterion for the affinity of cavitation. The cavitation specific speed C of pump can be deduced from the affinity theory of pump.

$$C=\frac{n\sqrt{Q}}{NPSH_r^{3/4}} \tag{6.16}$$

If the unit of n is r/min, the unit of Q is m³/s, and the unit of $NPSH_r$ is m, the above formula needs to be multiplied by a coefficient of 5.62, i. e.

$$C=5.62\,\frac{n\sqrt{Q}}{NPSH_r^{3/4}} \tag{6.17}$$

From the above formula, the larger the C value, the smaller the $NPSH_r$, the better the cavitation performance of the pump. Therefore, the value of C can easily distinguish the performance of pump cavitation. As with the specific speed, the discharge Q in the formula to calculate cavitation specific speed of the double suction pump, and the discharge is half.

Cavitation specific speed and specific speed of pump are similar criteria of pump. The specific speed n_s of the pump is the outlet affinity criterion of the pump, which is used for the energy performance conversion of the pump; while the specific speed C of the pump is the inlet affinity criterion of the pump, which is used for the cavitation performance conversion of the pump.

6.8　Pump installation height

The installation height H_x of the pump refers to the vertical height from the datum level of the pump (Fig. 6.8) to the water level of the intake sump, which is the basic basis for the design of each elevation of the pump station. H_x can be calculated by the permitted suction vacuum H_s or the $NPSH_r$ of the pump, while H_s and $NPSH_r$ are determined by the cavitation performance of the pump, which is usually marked on the pump sample.

6.8.1 Permitted suction vacuum head

The suction vacuum head of the pump refers to the value that the pressure at the inlet of the pump is lower than the atmospheric pressure. If the pressure at the liquid level of the water inlet sump is an atmospheric pressure, the vacuum head at the inlet of the pump is the difference between the atmospheric pressure and the pressure at the S—S section of the pump inlet. Usually a vacuum meter is installed at the S—S section. The reading on the vacuum meter is the vacuum head of the pump, use meter of water column height as unit:

$$H_s = \frac{P_a}{\rho g} - \frac{P_s}{\rho g} \tag{6.18}$$

Where, $\dfrac{P_s}{\rho g}$ is the pressure of S—S section at pump inlet.

The suction vacuum head of the pump is also a parameter reflecting the cavitation performance of the pump. The larger the suction vacuum head, the better the anti-cavitation ability of the pump.

For centrifugal pumps, the suction vacuum head value is generally used as a parameter to reflect the cavitation performance of the pump, and its value is determined by the test. For the sake of safety, the permissible value of the pump is usually set a little smaller when it is manufactured, that is, the suction vacuum height value of the pump just before cavitation occurs minus a safety amount (0.3 - 0.5m), which is called the allowable suction vacuum height of the pump, expressed as $[H_s]$.

6.8.2 Calculation of installation height of centrifugal pump and volute mixed flow pump

1. Calculate with $[H_s]$

Horizontal centrifugal pump and volute mixed flow pump, generally installed above the water surface, from the critical state of cavitation $[H_s]$:

$$[H_s] = H_x + h_x + \frac{v_s^2}{2g}$$

so:

$$H_x = [H_s] - h_x - \frac{v_s^2}{2g} \tag{6.19}$$

2. Calculate with $[NPSH_r]$

If the pump cavitation parameter provided is $[NPSH_r]$, according to the critical state

$$[NPSH_r] = \frac{P_a}{\rho g} - H_x - h_x - \frac{P_{cav}}{\rho g}$$

so:

$$H_x = \frac{P_a}{\rho g} - h_x - [NPSH_r] - \frac{P_{cav}}{\rho g} \tag{6.20}$$

Simplified at room temperature

$$H_x = 10.0 - h_x - [NPSH_r] \tag{6.21}$$

Because the safety margin has been considered in $[NPSH_r]$, the calculated H_x value does not need to consider taking the safety margin any more.

3. Altitude and vaporization pressure correction

It must be noted that the $[H_s]$ of the pump refers to the value when the altitude is 0 and the ambient temperature is 20℃ . If the place of use and temperature are non-standard, the $[H_s]$ should be corrected as follows

$$[H_s]'= [H_s] +\frac{P_a}{\rho g}-10.33-\frac{P_{cav}}{\rho g}+0.24 \tag{6.22}$$

Where, $[H_s]'$ is the allowable vacuum height after altitude and temperature correction, m; $[H_s]$ is the allowable suction vacuum height under standard condition, m; $\dfrac{P_a}{\rho g}$ is the atmospheric pressure at the pump installation site, m, as shown in Table 6.2; $\dfrac{P_{cav}}{\rho g}$ is the vaporization pressure at working temperature, m, as shown in Table 6.1.

If the H_x calculated in the formula above is greater than 0, it means that the reference surface of the pump is above the water level, indicating that the pump has positive suction range. If H_x is less than 0, it means that the reference surface of the pump must be below the water surface, i. e. the pump impeller must be submerged and the suction distance of the pump must be negative.

6.8.3　Calculation of H_x of axial flow pump and guide vane mixed flow pump

Due to the short inlet pipeline of the axial flow pump and the guide vane mixed flow pump, only one bellmouth in small and medium-sized pumping stations, the loss of intake can be neglected.

$$H_x=10.0- [NPSH_r] \tag{6.23}$$

In order to facilitate the start-up, even if the calculated H_x value is positive, the impeller is not installed above the water surface. The depth of the impeller datum from the water surface is generally 1.2 times the diameter of the bellmouth or not less than 0.5 –1m.

6.9　Causes of pump cavitation and protective measures

Cavitation is unavoidable in operation of pump station. The main reasons including:

(1) The pump is more likely to operate under non-design conditions.

(2) Even if the pumping station suction system is reasonably designed, cavitation may still occur in some parts, such as clearance cavitation at blade and pump casing, orifice ring and volute.

(3) Abrasion of pump by sediment-laden water source aggravates vaporization and cavitation damage.

(4) The high speed and large discharge of the pump bring about the severity of cavitation.

(5) The improper selection of pumps in pumping stations leads to the long-term deviation of pumps from the design conditions.

To prevent water pump cavitation, measures can be taken from two aspects: improving the pump and using conditions.

(1) Improve the anti-cavitation ability of the pump and reduce the necessary residual cavitation of the pump

1) Improve the design and manufacturing process of the pump.

2) Surface protection technology can be used to repair the overflow parts which have been damaged by cavitation. That is, the use of materials with a certain degree of hardness and toughness, or worse texture parts of the surface strengthening treatment, so that the surface hardness and toughness increased. Surface protection technology can significantly improve the anti-cavitation and abrasion ability of pump blades and impeller casings, and prolong their service life.

3) For important pumps, stainless steel and other anti-cavitation materials should be used to make pumps.

(2) Improve pump suction device and operating conditions, and increase effective cavitation margin of the device

1) When arranging the water inlet pipeline of the pump, it is necessary to minimize unnecessary pipeline accessories, such as the inlet bottom valve, elbow, gate valve, etc. The pipeline should not be too long, the inlet diameter can be selected to be larger, so as to reduce the discharge in the pipeline and reduce the head loss.

2) The inlet sump and diversion structure of the pumping station should be designed reasonably to make the water flow smoothly into the pump, avoid bad flow patterns such as bias current, backflow, vortex, so that the inlet water flow of the pumping station is entrained with air. If necessary, model test should be carried out and necessary anti vortex measures should be taken to improve the inlet flow pattern.

(3) The trash racks should be cleared in time to avoid excessive water level drop in the intake water of pumping station.

(4) If necessary, under the premise of meeting the operation requirements, the pump speed can be reduced appropriately, because the residual cavitation of the pump is proportional to the square of the speed ratio.

(5) When the cavitation occurs, small amount of airs can be added to the suction side in an emergency manner to reduce the vacuum in the cavitation area, but the amount and duration of air supply must be strictly controlled.

第7章 泵站工程规划

灌、排泵站工程规划是区域水利规划的一部分,城市给、排水泵站工程规划是市政工程规划的一部分。因此,泵站工程规划应在区域或市政工程规划的基础上进行,它不仅关系到工程本身,而且关系总体规划和长远规划的合理性与经济性,关系到工程的投资、效益和运行管理。合理的工程规划不仅在工程兴建时达到减少工程量,节省投资,而且有利于工程建成后的运行管理,为降低工程运行成本创造有利条件。泵站工程规划应在流域或地区水利规划的基础上,根据全面规划、综合治理、合理布局、经济可行的原则,正确处理好近期与远期、整体与局部的关系,协调好与其他用水部门的关系,使水泵站工程发挥最大效益。

泵站工程规划的主要任务如下:

(1)划分灌溉区(排水区)。

(2)确定泵站的规模、设计标准、设计参数等。

(3)确定泵站工程的组成(泵站站址、枢纽布置、运行方案等)。

7.1 泵站等别

泵站的等级和规模应根据工程任务,应考虑近期目标为主,并考虑远景发展要求,综合分析确定。目前,根据《泵站设计规范》(GB 50265—2010)我国将泵站工程分为5级,见表7.1。对工业、城镇供水泵站等级的划分,应根据供水对象、供水规模和重要性确定。泵站等别决定了其主要建筑物、次要建筑物和临时性建筑物的级别。

由多级或多座泵站联合组成的泵站工程的等别,可按其整个系统的分等指标确定。当泵站按分等指标分属不同等别时,应以其中的高等别为准。

表 7.1　　　　　　　　　　　　　泵 站 等 别
Table 7.1　　　　　　　　　　　Pumping station grade

泵站等别 pumping station grade	泵站规模 pumping station scale	灌溉、排水泵站 irrigation and drainage pumping stations		工业、城镇供水泵站 industrial and urban water supply pumping stations
		装机流量 discharge/（m³/s）	装机功率 power/万 kW	
I	大（1）型 large scale, type I	≥200	≥3	特别重要 particularly importance
II	大（2）型 large scale, type II	200~50	3~1	重要 importance
III	中型 medium scale	50~10	1~0.1	中等 medium importance

续表

泵站等别 pumping station grade	泵站规模 pumping station scale	灌溉、排水泵站 irrigation and drainage pumping stations		工业、城镇供水泵站 industrial and urban water supply pumping stations
		装机流量 discharge/（m³/s）	装机功率 power/万 kW	
Ⅳ	小（1）型 small scale, type Ⅰ	10～2	0.1～0.01	一般 general importance
Ⅴ	小（2）型 small scale, type Ⅱ	<2	<0.01	—

泵站建筑物应根据泵站所属等别及其在泵站中的作用和重要性分级，其级别按照表7.2确定。

表 7.2　　　　　　　　　泵站建筑物级别划分

Table 7.2　　　　　　　　**Rank of pumping station buildings**

泵站等别 pumping station grade	永久性建筑物级别 permanent building rank		临时性建筑物级别 temporary building rank
	主要建筑物 main buildings	次要建筑物 secondary buildings	
Ⅰ	1	3	4
Ⅱ	2	3	4
Ⅲ	3	4	5
Ⅳ	4	5	5
Ⅴ	5	5	—

7.2　灌区、排水区的分区、分级

7.2.1　灌区分区

灌区平面区域划分应按照最高经济效益的原则确定。根据泵站建设经验，灌区平面划分有以下几种基本形式。

（1）一站提水、一区灌溉。全灌区只建一座泵站，由一条干渠控制全部灌溉面积。泵站将水提升到灌区的制高点，然后由渠系向全灌区供水。此方案适用于面积较小、扬程较低、地面高差不大、输水渠道不长的小型灌区，如图 7.1 所示。地形高差不大的小型灌区大多采用这种布置方式。

（2）多站提水、分区灌溉。灌区由于地形或沟河分割，如采用一站提水、一区灌溉的方案，势必增加渠系河交叉建筑物工程量，以致增加工程投资，此时以分散建站分区灌溉为宜，如图 7.2 所示。

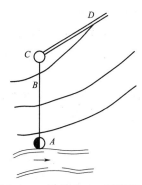

图 7.1　一站提水、一区灌溉

Fig. 7.1　Water lifting in one station and irrigation in one area

A—泵站 pumping station；B—出水管道 outlet pipe；

C—渠首 headworks；D—干渠 main canal

（3）多站分级提水、分区灌溉。在扬程较高、地形变化较大的灌区，为避免出现提升到高处的水再流回低处灌溉，造成能源浪费，可将全区分为若干高程不同的灌区，采用梯级泵站进行分级抽水，每级灌区由一条干渠控制，前一级站的抽水量除满足本灌区所需之外，还要供给后一级站的抽水量。这种方式适于地面高差较大或地形上有明显台地的地区，如图 7.3 所示。由于灌区本身是一个水、机、电综合的工程系统，需应用系统工程优化方法来确定梯级泵站的分级分区问题。

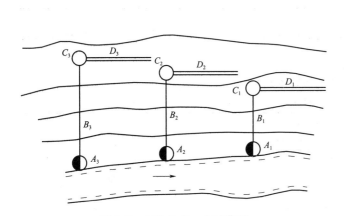
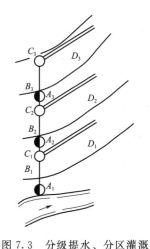

图 7.2　多站提水、分区灌溉

Fig. 7.2　Multi station pumping and
district irrigation

A—泵站 pumping station；B—出水管道 outlet pipe；
C—渠首 headworks；D—干渠 main canal

图 7.3　分级提水、分区灌溉

Fig. 7.3　Grading water lifting
and zonal irrigation

A—泵站 pumping station；B—出水管道 outlet pipe；
C—渠首 headworks；D—干渠 main canal

（4）一站分级提水、分区灌溉。对于某些地面高差较大但是面积不大的灌区，可采用一站分级提水、分区灌溉，即在同一泵站安装几台不同扬程的水泵，分高、低两个出水池和相应的渠道供水。

7.2.2　排水区分区、分级

排水区的分级、分区要尽可能满足内外水分开、高低水分开，并充分利用自流排水的条件。

内外水分开主要是洪涝分开，避免外河洪水侵入圩内；其次在排水区内部还要求内河和农田分开。即排涝时既要利用外河、内河、农田的滞蓄能力，又要建闸、筑堤、分级控制。

高低水分开就是要求等高截流、高水高排，低水低排，避免高水汇集到低地而加长排水时间，产生或加重涝害；同时，实现高水高排，也是避免高水汇集到低地增大排涝扬程，利于减少排涝站装机容量和运行费用。另外，由于高水往往有自排的可能，应充分利用时机，及时自流排水。

1. 排水区划分

（1）沿江滨湖圩内排水区的划分。湖区、圩区地面虽然平坦，但也会有一定高差，尤其是面积较大的地区，各处自流外排的条件不同，须统一规划，进行分区排水。分区时，应根据地形特点、承泄区水位条件，适当兼顾原有排水系统。对地势较高、有自排条件的地区，尽量利用自排条件，划分为高排区；对地势较低、排涝期间外水位长期高出田面的地区，划分为低排区，以提排为主；介于上述两者之间的地区，可采取自排与提排相结合的排水方式。

（2）半山半圩地区排水区的划分。这类地区后临丘陵地区或高地，前沿江、湖，俗称半山半圩地区。由于汛期外水位高于圩内农田，同时高处客水又流向下游，因此易成涝灾。分区时，要在山圩分界处大致沿承泄区设计外水位高程的等高线开挖截流沟或撤洪沟。使山、圩分排，高低分排，减少泵站的装机容量。

（3）滨海和感潮河道地区排水区的划分。对于滨海和感潮河段地区受洪水影响较小而潮汐影响较大的排水系统，可按地区高程划分排水区。地面高于平均感潮潮位者为畅排区，低于平均低潮位者为非畅排区，介于上述两者之间的为半畅排区。畅排区可以自流排水，非畅排区依靠泵站提排，而半畅排区则应考虑增加排水出口，缩短排水路径，并于出口建挡潮闸，利用落潮间歇自流抢排。若这样处理后仍不能满足排水要求，考虑建外排（涝）站，在涨潮期间闭闸提排。这类地区的规划，应详细分析涝水、洪水和潮位的关系，以调整不合理的排水分区和排水出口，尽可能扩大畅排区和缩小非畅排区，减少排涝站的装机容量和提排费用。

2. 排水区分级

排水区分级包括一级排水、二级排水和一、二级排水结合三种方式。

（1）一级排水就是由排涝站直接将涝水排入承泄区，如图7.4（a）所示；或由排涝站将涝水先排入蓄涝区，而蓄涝区的涝水则待外水位降低时再开闸自排，如图7.4（b）所示。一级排水方式适用于排水面积不大、装机容量较小及扬程不高的排水区。

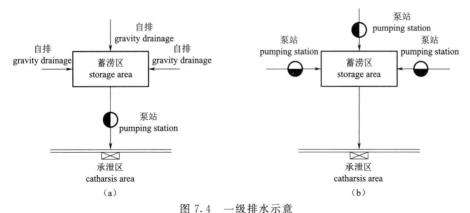

图7.4 一级排水示意

Fig. 7.4 One-stage drainage mode

(a) 方式一；(b) 方式二

(a) Mode one；(b) Mode two

（2）二级排水方式，即在低洼地区建小站，将涝水排入蓄涝区内，这种站称为二级站或内排站，一般排水扬程较低；而蓄涝区内的涝水则需要另外建站外排，这种站称为一级站或外排站,如图7.5所示。如蓄涝容积较大时,除利用泵站提排外,还可以利用蓄涝区滞蓄涝水,待外水位降低后再开闸排蓄涝区滞蓄的涝水,利用闸站配合排水,以减少外排站装机容量。二级排水方式适用于排水面积较大、地形比较复杂、高低不平及扬程较高的排水区。

拥有大型排涝泵站的排水区,由于控制范围大,涝区内地形复杂,应优先采用二级排水方式。这样,在布局上,集中与分散结合;在规格上,大、中、小结合。此外,二级站(内排站)一般容量小,运用灵活,能适应局部低地排水或低地要排、高地要灌的需要。

滨海和感潮河段的排水区,如可利用退潮排水,而蓄涝容积又较大时,则内排站可先排低地涝水,待退潮后再开闸排蓄涝区。这种情况下内排站起外排站的作用,属一级排水方式。若蓄涝容积较小或退潮后外水位仍比较高,不能完全依靠开闸自排,则应另设外排站配合提排涝水,这就是二级排水方式。

(3)实际工程中,上述两种排水方式往往不是截然分开,有时外排站既可排蓄涝区的涝水,又可直接排出,在运用上采取先排田后排蓄涝区。当泵站排田时为一级排水,而排蓄涝区时则为二级排水,如图7.6所示。

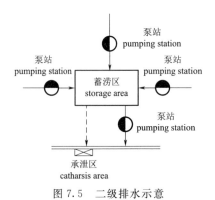

图7.5 二级排水示意

Fig.7.5 Two-stage drainage mode

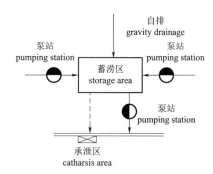

图7.6 一、二级结合排水方式

Fig.7.6 One and two combined drainage methods

排水方式的选择是排水站规划中的一项重要工作,应通过详细技术经济比较确定,此外,排涝站的规划还要考虑综合利用。根据需要,可以通过合理布置泵站建筑物,把一座排涝站设计成既能提排提灌,又能自排自灌,一站多用,扩大工程效益。

7.2.3 站址选择

站址选择应和灌区、排水区的划分同时进行。分区分级确定后,选择具体站址还应考虑以下各因素。

(1)地形。站址地形要平坦、开阔,以利泵站建筑物的总体布置。泵站应能控制全区面积,灌溉泵站站址选在全区上游地势较高地段,排水泵站站址选在排水区内地势较低、距离排水干河出口处较近处。

(2)地质。站址应选在坚固的地基上。泵房及进水建筑物常建在较深的开挖基面上,应避开淤泥、流沙、湿陷性黄土、膨胀土等地基,以减少地基处理的费用。

(3)水源。应选择水量充沛、水质好的河段和湖泊、水库作为水源。对于河流,应尽量选在河段顺直、河床稳定处。如遇弯曲河段，应选在坡陡、泥沙不易淤积的凹岸，并避免在有浅滩、支流汇入和分岔的河段建站。排涝站还应尽可能选择在外河水位较低的地段，以便降低排涝扬程。

（4）综合利用。排灌结合泵站的站址要能为抽水排灌与自流排灌相结合创造条件，还可为其他用途提供场地。

（5）电源。站址应尽可能靠近电网，以减少输、变电工程投资。

（6）其他。站址处交通应方便，并应靠近乡镇、居民点，以利设备、材料的运输和建设、运行管理。

7.3 泵站设计参数的确定

泵站设计参数主要是设计流量和设计扬程，它们是水泵选型和泵站建筑物设计的依据。设计参数影响设备投资、泵站规模以及泵站的效益。因此，在确定这些参数时要进行充分的论证。

7.3.1 泵站流量

1. 灌溉泵站设计流量的确定

灌溉泵站设计流量应在满足一定的灌溉设计标准下，根据作物的灌溉制度、灌水模数、灌溉面积、渠系水利用系数及灌区内调蓄容积等综合分析计算确定。灌溉设计标准是反映灌溉水源或泵站提水能力对于农田灌溉用水保证程度的一项指标，是确定泵站工程的规模和设计参数的重要依据，应根据灌区的水源状况、作物布局、水文气象条件、地形、土质、当地农业生产的规划及经济条件等因素认真分析决定。灌溉设计标准一般用灌溉设计保证率或抗旱天数作为设计标准。

2. 排水泵站设计流量的确定

排水泵站设计流量取决于泵站的任务。平原圩区的排水泵站一般以排为主，结合灌溉，而排水又以排涝为主，兼顾排渍和降低地下水位。因此应以排涝流量作为排水泵站的设计流量。

排涝流量是根据一定的排涝设计标准推求的。如设计标准过低，则减轻涝灾的作用不大，会影响农业的生产，也影响群众的生活；设计标准过高，则工程规模过大，在一般年份泵站不能充分发挥作用，造成浪费。因此，排涝设计标准应根据各地区农业和国民经济各行各业发展需要和水利设施的现状，并根据规划期人力、物力、财力等条件，综合分析确定。

推算排涝设计流量一般采用排水模数法、平均排除法和调蓄演算法。排水天数主要根据作物的允许耐淹时间决定。对于小面积常采用1日暴雨2～3天排除；对于大排水面积常采用3日暴雨4～5天排除。对于旱作物地区，排水时间应较水田地区短。

7.3.2 泵站扬程

1. 灌溉泵站扬程计算

泵站上下游水位有出水池水位（上游）和进水池水位（下游），上游水位决定于灌区田面高程及灌溉流量等因素，下游水位决定于水源水位。当灌区渠首高程已确定，可根据泵站进、出水池特征水位计算泵站扬程。

（1）进水池水位。进水池水位同样又分最高水位、设计水位、最高运行水位、最低运行水位、平均水位等，对无引渠或引渠较短的泵站，其进水池水位确定方法见表7.3。如引渠较长，进水池相应水位确定，应扣除从水源（取水口）至进水池的水力损失。从河床不稳定的河道取水时，还应考虑河床变化对水位的影响。

表 7.3　　　　　　　　　　灌溉泵站进出水池水位确定方法
Table 7.3　　**Water level of inlet and outlet sumps of irrigation pumping station**

特征水位 characteristic water level	进水池（水源） inlet sump（water source）	出水池 outlet sump
最高水位 maximum water level	根据建筑物防洪标准确定 determine according to building flood control standard	当出水池接输水河道时，取输水河道的校核洪水位；当出水池接输水渠道时，取与泵站最大流量相应的水位 When the outlet sump is connected with the water delivery channel, take the flood level of the water delivery channel; when the outlet sump is connected with the water delivery channel, take the water level corresponding to the maximum flow of the pumping station
设计水位 design water level	从河流、湖泊或水库取水时，取历年灌溉期水源保证率为85%～95%的日平均或旬平均水位；从渠道取水时，取渠道通过设计流量时的水位 When drawing water from rivers, lakes or reservoirs, the daily or ten-day average water level with guaranteed rate of 85%-95% in irrigation period over the years is taken; when drawing water from channels, the water level when taking channels through design flow is taken	取按灌溉设计流量和灌区控制高程的要求推算的出水池的水位 The water level of the outlet sump is calculated according to the design discharge of irrigation and the control elevation of irrigation area
最高运行水位 maximum operating water level	从河流、湖泊取水时，取重现期5～10年一遇洪水的日平均水位；从水库取水时，根据水库调蓄性能确定；从渠道取水时，取渠道通过加大流量时的水位 When fetching water from rivers and lakes, the average daily water level of once-in-10-year flood with a return period of 5 to 10 years is taken; when fetching water from reservoir, the water level is determined according to reservoir storage performance; when fetching water from channel, the water level is taken through increasing discharge	取与泵站加大流量相应的水位 Take the water level corresponding to the increased flow of the pumping station
最低运行水位 minimum operating water level	受潮汐影响的泵站，其最低运行水位取历年灌溉期水源保证率为95%～97%的日最低潮水位 For the pumping station affected by the tide, the lowest operating water level is taken from the daily lowest tidal water level with the guaranteed rate of water source of 95%-97% in the irrigation period over the years	取与泵站单泵流量相应的水位；有通航要求的输水河道，取最低通航水位 Take the water level corresponding to the single pump discharge of the pumping station and the lowest navigation level for the water conveyance channel with navigation requirements

续表

特征水位 characteristic water level	进水池（水源） inlet sump（water source）	出水池 outlet sump
平均水位 average water level	从河流、湖泊或水库取水时，取灌溉期水源多年日平均水位；从渠道取水时，取渠道通过平均流量时的水位 When drawing water from rivers, lakes or reservoirs, the average daily water level of water sources in irrigation period for many years shall be taken; when drawing water from channels, the water level when the channels pass through the average flow shall be taken	取灌溉期多年日平均水位 Average daily water level for many years during irrigation period is taken

1）进口最高运行水位。用于校核水泵及配套动力机的工作状况，也用于校核工程安全稳定性。

2）进口设计水位。用以确定泵站设计扬程。

3）进口最低运行水位。用以确定水泵的安装高程和校核工程的安全。

4）进口最高水位。对于泵房不直接挡水的泵站，用以确定机房地面或电机层楼板高程以及设计挡水墙，校核工程的抗浮、抗滑稳定性。当泵站进口出现最高水位时，泵机组不一定需要运行，因此作为确定最高水位数值的标准不是灌溉保证率，而是泵站建筑物的等级。

5）进口最低水位。根据水源枯水位资料以泵站建筑物等级所要求的标准用频率分析确定。设计时，用以校核工程安全。

（2）出水池水位。灌溉泵站的出水池水位是灌溉渠系由渠尾到渠首逐级推算出的灌溉干渠的渠首水位，其确定方法见表 7.3。

（3）灌溉泵站特征扬程的确定。根据上述出水池水位和进水池水位差组合，可以算出泵站的各种扬程：

设计扬程＝出口设计水位－进口设计水位

最高扬程＝出口最高水位－进口最低运行水位

最低扬程＝出口最低水位－进口最高运行水位

在设计扬程下，应保证泵站设计流量的要求。在平均扬程下，水泵应在高效区工作。平均扬程可按式（7.1）计算加权平均扬程，或按泵站进、出水池平均水位差计算。

$$H = \frac{\sum H_i Q_i t_i}{\sum Q_i t_i} \tag{7.1}$$

式中：H 为加权平均扬程，m；H_i 为第 i 时段泵站进、出水池运行水位差，m；Q_i 为第 i 时段泵站提水流量，m^3/s；t_i 为第 i 时段历时，d。

2. 排水泵站扬程计算

（1）内、外河水位。排水站的内河水位一般有内河最高水位、内河最高运行水位、内河设计水位、内河最低运行水位等，其中内河最低运行水位与最高运行水位又分别称起排水位与停排水位。

排水站的外河水位一般有防洪水位、外河最高运行水位、外河设计水位、外河最低运行水位等，其在泵站设计中的作用与灌溉泵站各水位的作用相似，各水位确定方法见表 7.4。

表 7.4　　　　　　　　　　　　排水泵站内、外河水位确定方法

Table 7.4　　　　　　　　**Calculation method of inland river and outer river**

特征水位 characteristic water level	内河 inland river	外河 outer river
防洪水位 flood control level		根据建筑物防洪标准确定 Determine according to flood control standard of buildings
最高水位 maximum water level	取排水区建成后重现期 10～20 年一遇的内涝水位 Waterlogging water level occurs once in 10 - 20 years after completion of intake and drainage area	
设计水位 design water level	取由排水区设计排涝水位推算到站前的水位；对有集中调蓄区或内排站联合运行的泵站，取由调蓄区设计水位或内排站出水池设计水位推算到站前的水位 The water level before the station is calculated from the designed drainage water level in the drainage area; for pumping stations with centralized storage area or combined operation of internal drainage station, the water level before the station is calculated from the designed water level in the reservoir area or the designed water level in the outlet pool of the internal drainage station	取承泄区重现期 5～10 年一遇洪水的 3～5 日平均水位；当承泄区为感潮河段时，取重现期 5～10 年一遇的 3～5 日平均潮水位 The average 3 - 5 days water level of once-in-5 - 10-year flood in the drainage reception district is taken; when the drainage reception district is a tidal reach, the average 3 - 5 day water level once-in-5 - 10-year flood level is taken
最高运行水位 maximum operating water level	取按排水区允许最高涝水位的要求推算到站前的水位；对有集中调蓄区或内排站联合运行的泵站，取由调蓄区最高调蓄水位或内排站出水池最高运行水位推算到站前的水位 The water level before the station is calculated according to the requirement of the highest allowable flood water level in the drainage area; for pumping stations with centralized regulating and storage area or combined operation of inland drainage station, the water level before the station is calculated from the highest regulating and storage water level in the reservoir area or the highest operating water level in the outlet pool of inland drainage station	当承泄区水位变化幅度较小，水泵在设计洪水位能正常运行时，取设计洪水位；当承泄区水位变化幅度较大时，取重现期 10～20 年一遇洪水的 3～5 日平均水位；当承泄区为感潮河段时，取重现期 10～20 年一遇的 3～5 日平均潮水位 When the water level in the drainage reception district changes slightly and the pump can operate normally, the design flood level is taken; when the water level in the drainage reception district changes greatly, the average 3 - 5 days water level of the once-in-10 - 20-year flood is taken; when the drainage reception district is a tidal reach, the average 3 - 5 days water level of the once-in-10 - 20-year flood is taken

续表

特征水位 characteristic water level	内河 inland river	外河 outer river
最低运行水位 minimum operating water level	取按降低地下水埋深或调蓄区允许最低水位的要求推算到站前的水位 The water level before the station is calculated according to the requirement of lowering groundwater depth or lowest allowable water level in the reservoir area	取承泄区历年排水期最低水位或最低潮水位的平均值 The average value of the lowest water level or the lowest tidal level in the drainage period over the years in the drainage reception area is taken
平均水位 average water level	取与设计水位相同的水位 Take the same water level as the design water level	取承泄区排水期多年日平均水位或多年日平均潮水位 Multi-year daily average water level or multi-year daily average tidal level in drainage period of drainage reception area

（2）排水泵站特征扬程的确定。已知内、外河水位，则可根据水位机遇组合得各种排涝扬程。

$$设计扬程＝设计外水位－设计内水位$$

如果最高运行外水位与最低运行内水位有相遇的可能，则最大扬程按下式计算

$$最大扬程＝最高运行外水位－最低运行内水位$$

如果最高运行外水位与最低运行内水位无相遇的可能性，可分别取下述两种水位组合，以其中较大者作为采用的最大扬程

$$最大扬程＝最高外水位－设计内水位$$

或 $$最大扬程＝设计外水位－最低运行内水位$$

$$最低扬程＝最低运行内水位－最高运行外水位$$

7.4 泵 站 机 组 选 型

水泵与配套的动力机是泵站的主要设备，称为主机组。所有辅助设备和水工建筑物均为主机组的运行服务。因此，主机组选型配套是否合理，直接影响泵站效益、土建工程和机电设备投资、设备利用率等，是规划中的重要问题。

7.4.1 水泵选型

1. 选型要求

水泵选型应满足以下要求：能符合一定设计标准下的供排水及灌溉设计流量与设计扬程要求；水泵运行在高效区；机组大小和台数应使泵站投资较省、便于维修和管理，运行费用较少；水泵稳定安全运行，汽蚀性能良好。

2. 水泵类型比较与选择

泵站主水泵型号主要是根据扬程和流量选定的，其次还要考虑现有泵类产品的供应情况。对于高扬程泵站基本选用双吸式离心泵或多级离心泵，其单机容量大小则由台数控制。对于中、低扬程泵站，有时有较多泵型可供选择，故在选型时，应对结构型式、安装

方式等进行技术经济比较。

各种水泵性能规格的选择图表有泵样本及水泵系列综合型谱图等，各个水泵厂也有产品目录可供查阅。

3. 选型步骤

（1）根据供（灌）排水区的规划要求，确定建泵站的流量和扬程这两个参数，一般由供（灌）、排水规划确定。根据水利规划确定泵站上下游水位、泵站设计总流量等关键参数。

（2）初选水泵台数与型号。

1）确定水泵台数。水泵台数的多少，决定水泵尺寸的大小和土建工程投资，因此需要高度重视。台数太少，保证率过低；台数过多，运行成本高，不便管理。从经验来讲，一般以 4～8 台为宜，最好不超过 8～10 台。确定机组台数后，可计算单泵设计流量 $Q_{站}/n_{台}$。

2）泵站净扬程计算。通过规划给定的上下游水位，计算泵站净扬程，包括设计净扬程、最高净扬程、最低净扬程等特征扬程。

3）管路损失扬程估算。由于此时水泵型号尚未选定，泵站进出水管路尚未设计布置，初选水泵时管路损失扬程采用估算法。高扬程泵站水泵采用等流量加大扬程的水泵选型方法，即估算的水泵扬程为设计流量下的净扬程与估算损失扬程之和。管路损失扬程宜取净扬程的 20%～30%。工程实践，尤其是南水北调东线一期泵站水泵选型经验显示，低扬程水泵装置的最优工况点向小流量偏离较大，相应的高效点装置扬程和水泵扬程接近，而水泵流量可适当增加，可取 10% 左右。采用等扬程加大流量的低扬程水泵选型方法，更加准确，可避免实际运行中长期偏低扬程大流量工况运行带来的各种运行问题。

4）初选泵型。根据单泵设计流量、初估水泵设计扬程，从水泵样本中的性能表、水泵综合性能曲线、水泵型谱图或从计算机数据库中选取 2 种以上泵型。根据水泵设计扬程、流量确定水泵转速或运行角度（图 7.7）。

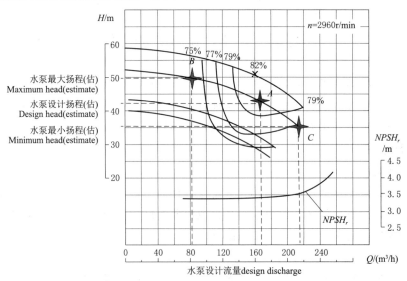

图 7.7 根据初估水泵特征扬程与设计流量选泵示意图（以离心泵为例）

Fig. 7.7 According to the preliminary estimation of pump characteristic head and design discharge, select pump schematic diagram (take centrifugal pump as an example)

（3）动力机配套。根据泵站出现的最不利工况选择动力配套功率，确定动力机型号。

（4）泵站进出水管路设计布置。

1）管路布置。根据水泵样本选取水泵泵型尺寸，布置进出水管路，选择合适的管路附件与断流设施。

2）泵站管路特性。分别计算管路局部、沿程损失，进而推算出泵站进出水管路阻力系数 S，绘制管路特性曲线。

3）泵站需要扬程特性。将各运行工况下的上下游水位差（泵站净扬程 H_{st}）加上管路损失（SQ^2），绘制各个特征净扬程下的需要扬程曲线。

（5）工况点校核。

1）确定各特征净扬程下水泵工况点。将水泵初选泵型的性能曲线（H-Q）和需要扬程曲线 H_r-Q 绘在同一张坐标中，两条曲线的交点即水泵的工况点 D、E、F（图 7.8）。

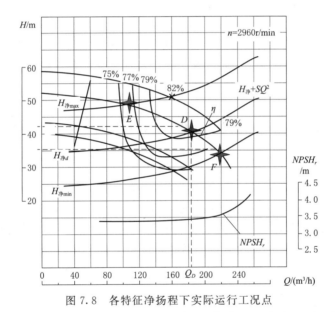

图 7.8 各特征净扬程下实际运行工况点

Fig. 7.8 Actual operating point under each characteristic net head

2）校核设计净扬程工况点（图 7.8 中 D 点）是否在所选泵型高效区，设计工况点的流量 Q_D 大于规划的单泵设计流量 $Q_{泵设计}＝Q_{泵站}$/机组台数。

3）校核最不利工况下（离心泵为最小扬程工况，图 7.8 中 F 点）动力机功率是否超载。

上述工况校核中如果有一条不满足要求，须重新选泵或更改进出水管路布置。2 个泵站水泵选型方案的性能参数重点比较设计工况点的效率和必需汽蚀余量。

4．选型应注意的问题

（1）台数。一般情况下，泵站流量小于 $1m^3/s$ 的场合可选 2 台泵，大于 $1m^3/s$ 的场合可考虑选 3～8 台泵，在供水保证率要求较高的场合，要考虑留有备用水泵，但总台数最多不宜超过 10 台。

备用机组数的确定应根据供水的重要性及年利用小时数，并应满足机组正常检修要求。对于重要的城市供水泵站，工作机组 5 台以下时，应增设 1 台备用机组，多于 5 台时，再增加 1 台备用机组；对于农用泵站，可适当减少备用机组，亦可参照有关规范。

（2）水泵安装形式。水泵安装形式一般有立式、斜式和卧式三种。卧式泵一般安装高程位于进口水面以上，开挖量小，安装要求比立式泵低，维修方便，工作条件好；但卧式机组占地面积大，一般启动前要抽真空。立式泵占地面积小，叶轮淹没在水面以下，无进水管路或进水管路短，启动方便；但其安装要求高，泵房高度大，工程造价较高。斜式泵介于立式泵和卧式泵之间。

（3）选用抗汽蚀性能好的水泵。选用抗汽蚀性能好的水泵可提高水泵安装高程，减少泵站开挖深度，节省工程投资。

（4）多因素综合考虑。水泵的选型，与泵站土建结构有直接的关系，常常需要和土建设计方案一起综合考虑，进行综合比较后决定。

7.4.2 动力机选型

水泵型号确定后还需为水泵选配合适的动力机及转动装置。驱动水泵最常用的是电动机，其次为柴油机。在购置水泵时，水泵生产厂商通常会配套供货，用户一般应进行计算复核或自行选配。

1. 配套功率计算

确定动力机配套功率时，按照水泵工作范围内最不利工况时的最大轴功率来计算。配套功率 N_{mt} 按下式计算

$$N_{mt} = K \frac{\rho g Q H}{1000 \eta_p \eta_{dr}} \quad (\text{kW}) \tag{7.2}$$

式中：Q 为水泵最不利工况时对应于最大轴功率的流量，m^3/s；H 为水泵最不利工况时对应于最大轴功率的扬程，m；η_p 为水泵最不利工况时对应于最大轴功率的效率；η_{dr} 为传动效率（0.9～1.0）。K 为动力机备用系数，按表 7.5 选取。

表 7.5　　　　　　　　　　　　　　　　备 用 系 数 K

Table 7.5　　　　　　　　　　　　Spare coefficient factor K

水泵轴功率 pump shaft power/kW	<5	5～10	10～50	50～100	>100
电动机 electric motor	2～1.3	1.3～1.15	1.15～1.1	1.1～1.05	1.05
柴油机 diesel engine		1.5～1.3	1.3～1.2	1.2～1.15	1.15

表 7.5 中的备用系数值，可按照小泵取大值、大泵取小值的原则选定，配套功率的确定，还要符合动力机的额定容量。

2. 电动机选型配套应注意的问题

电动机是电力泵站的主要设备之一，其选择正确与否将直接影响抽水装置的效率及泵站运行安全。电动机选型主要考虑以下几方面内容：

（1）电动机类型的选择。泵站中常采用的电动机有异步电动机和同步电动机。鼠笼型异步电动机结构简单、运行可靠、维护方便、价格低廉，且易于实现自动控制，因此使用较多。虽然启动电流较大，瞬间压降大，启动力矩小，但由于水泵属轻载启动，采用此类电动机已能满足工程要求。对选用大容量鼠笼型异步电动机常需配用启动设备进行降压启动。当电网容量不能满足鼠笼型电动机启动要求时，可考虑用绕线型电动机。

同步电动机虽然造价高、结构复杂、维护管理麻烦，但具有较高的效率和功率因数，还可作调相运行，向电网输送无功功率。同时具有较大的启动力矩，对电网电压波动适应性较强等特点，适用于大型泵站及使用时间较长的场合。

（2）电动机的安装形式。电动机的安装形式一般与水泵的安装形式一致，即卧式水泵配用卧式电动机，立式水泵配用立式电动机。工程中也有用卧式电机配立式水泵的，需采用间接传动机构。

（3）电动机电压的选择。按电动机的功率选择电源电压：功率在 200kW 以下的，选用 0.4kV 的三相交流电；功率在 300kW 以上的，选用 10kV（或 6kV）的三相交流电；功率在 200～300kW 之间的，两种电压均可选，可结合当地电网条件及技术经济比较后合理确定。

（4）电动机转速的选择。与水泵配套的电动机其转速必须满足水泵转速的要求，如果水泵转速与电动机样本上的额定转速差小于 2%，可以直接采用直联方式，否则要采用间接传动装置。

7.5 泵 站 枢 纽 布 置

泵站枢纽布置是泵站所有建筑物的总体布置，包括：取水、引水、进水建筑物；泵房；变电站，出水建筑物等相互之间的位置和布置形式。泵站枢纽布置对整个工程的造价以及泵站的运行管理等都有很大影响，究竟采用何种布置形式，都要根据排灌任务要求，地形、水系条件、原工程情况，拟选泵机组的性能和结构特点等综合考虑，以求得经济合理的方案。

泵站进水建筑物包括前池、进水池和进水管道等。对于有引渠的灌溉泵站，引水渠亦属于进水建筑物；对建在多泥沙河流上的泵站，前池部分往往还建有沉沙池。出水建筑物包括出水管道（流道）、出水池等。泵站站房（泵房）是安装水泵、动力机及其辅助设备以及泵站附属设备的建筑物，是泵站各建筑物中的主体工程。

7.5.1 枢纽布置应考虑的问题

影响枢纽布置的主要因素是建站的目的、水源的种类、水位变幅和水质，以及建站地点的地形、地质及水文等条件。

（1）枢纽布置必须服从区域治理规划要求，根据站址情况并参考已建泵站的经验，进行方案比较。

（2）要创造良好的水流条件，引水、进水流态要平稳，避免产生回流、死水区和水流漩涡。

（3）能充分发挥各个组成建筑物的作用，即尽可能以最少的建筑物发挥最大的效益。

（4）各个组成建筑物类型的选择和相互关系，要保证运行安全，管理方便，还要为泵站的扩建留有余地。

（5）尽量减少开挖、占用农田及原有建筑物的拆迁。

（6）尽可能考虑综合利用，如排灌结合，抽引与自流结合，泵站与交通结合等。

7.5.2 灌溉泵站枢纽布置形式

1. 有引水建筑物的泵站布置形式

图7.9所示为有引水明渠的泵站枢纽布置，适用于从江河、湖泊中取水，岸边坡度较缓，水源水位变幅不大的站址。

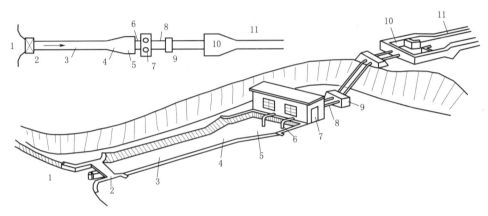

图7.9 有引水明渠泵站枢纽

Fig.7.9 Pumping station junction with diversion channel

1—水源 water source；2—进水闸 intake sluice；3—引水渠 diversion channel；4—前池 forebay；5—进水池 inlet sump；
6—进水管 intake pipe；7—泵房 pump house；8—出水管 outlet pipe；9—镇墩 anchor block；
10—出水池 outlet sump；11—灌溉干渠 main irrigation channel

2. 无引水建筑物的泵站布置形式

当水源岸边较陡，水位变幅不大，或灌区距水源较近时，常将泵站与取水建筑物合并，直接建在水源岸边或水源中（图7.10）。因此，泵房本身需要挡水。这种形式的优点是取消了引水建筑物，减少了维护工作量；缺点是泵房结构复杂，需要考虑防洪安全。

7.5.3 排水泵站枢纽布置形式

单纯排涝的泵站枢纽布置形式通常如图7.11所示，出水池紧靠河堤，出水池的出口与泄水涵洞相连，通过水泵提升到出水池的

图7.10 无引水明渠泵站枢纽

Fig.7.10 Pumping station junction without diversion channel

涝水经泄水涵洞泄入外河。也有不用泄水涵洞而用明渠泄水。泄水闸可起防洪作用，当外河水位高而又不需要提水排涝时，泄水闸关闭。

根据泵房与排水闸的相互关系，分为分建式和合建式两种布置形式。

（1）分建式。泵房与排水闸分开建造，引渠从排水河中取水。这种形式适用于原先有排水闸，而单靠自流排水无法解决内涝问题，需要建泵站在关闸期间排水。

（2）合建式。泵房与排水闸合建在一起，当内河水位高于外河时，开闸排水，当外河水位高于内河时，关闸开机排水。

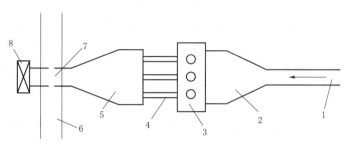

图 7.11　排涝泵站枢纽示意

Fig. 7.11　Drainage pumping station junction

1—排水干沟 drainage main ditch；2—前池 forebay；3—泵房 pump house；4—出水管 outlet pipe；

5—出水池 outlet sump；6—河堤 river embankment；7—泄水涵洞 discharge culvert；

8—泄水闸 discharge sluice

7.5.4　排灌结合泵站枢纽布置形式

1. 一站四闸站布置

对于既有灌溉任务，又有排涝任务的排灌结合的泵站工程，配套建筑物相应较多，更要讲究合理布置。为合理布置，首先应当确定必不可少的配套建筑物及其调节运用关系。图 7.12 所示的泵站枢纽布置称一站四闸布置，由四座节制闸和一座泵站组成，这种布置能满足提灌、提排、自引、自排需要。

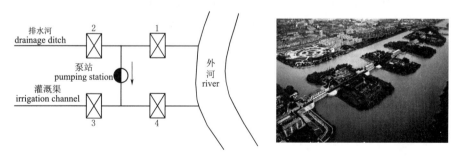

图 7.12　灌排结合泵站枢纽布置

Fig. 7.12　Layout of irrigation and drainage combined pumping station

提水灌溉时，开闸门 1、3，关闸门 2、4。

提水排涝时，开闸门 2、4，关闸门 1、3。

自流引水时，开闸门 3、4。

自流排水时，开闸门 1、2。

一站四闸布置形式的优点是进水条件好，排灌调度方便，矛盾较少；其缺点是水工建筑物较多，土建投资较高。

2. 双向流道闸站结合形式

排灌结合泵站的枢纽布置形式很多，往往与沟渠的走向有很大关系。在大型泵站建设中，为满足自引、自排及双向提水需要，常常将流道设计成 X 形，即双向流道闸站结合形式，如图 7.13 所示。站房下层既是进水流道，又可作为引水或排水的涵洞，上层是出水流道，进水和出水都为双向。如外河水位较低，可以打开闸门 2 和 4，内河水则可自流排入外河；若外河水位较高不能自排，则关闭闸门 1、4，打开闸门 2、3，可提排入外河。灌溉时，若外河水位较高，可开闸门 2 和 4，自引外河水灌溉；若外河水位较低，则关闭闸门 2、3，打开闸门 1、4，进行提水灌溉。这种泵站的站房直接挡水，为堤身式泵站，适用于扬程较低、内外水位变化幅度不大的场合。

与一站四闸布置形式相比，双向流道闸站结合布置占地面积小，工程投资省，便于集中管理，具有结构简单、双向抽水、抽引结合、运行效率较高等优点，因而得到了较广泛的应用。

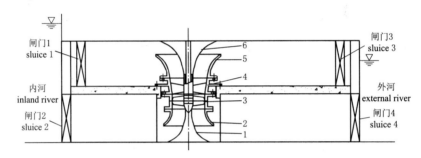

图 7.13　箱涵式双向流道闸站结合泵站剖面图

Fig. 7.13　Profile of combined pumping station with box-culvert two-way flow passage

1—进口导水锥 inlet cone；2—吸水喇叭 suction bellmouth；3—转轮室 casing；

4—导叶体 guide vane；5—出水喇叭 outlet bellmouth；6—出口导水锥 outlet cone

Chapter 7　Planning of Pumping Station Projects

The planning of irrigation and drainage pumping stations is a part of regional water conservancy planning, and the planning of urban water supply and drainage pumping stations is a part of municipal engineering planning. Therefore, the planning of pumping station project should be carried out on the basis of regional or municipal engineering planning. It is not only related to the project itself, but also to the rationality and economy of the overall plan and long-term plan, and related to the investment, benefit and operation management of the project. Reasonable project planning not only reduces the amount of project and saves investment during the construction of the project, but also facilitates the operation and management after the project is completed, and creates favorable conditions for reducing the operation cost of the project. On the basis of comprehensive planning, comprehensive treatment, rational layout and economic feasibility, the planning of pumping station projects should correctly handle the relationship between short-term and long-term, overall and partial, coordinate with other water departments, and bring the pumping station projects into full play.

The main tasks of the pumping station project planning are:

(1) Zone the irrigation district (drainage district).

(2) Determine the scale, design standard and design parameters of the pumping station.

(3) Determine the composition of the pumping station project (pumping station site, junction layout, operation scheme, etc.).

7.1　Pumping station grade

The grade and scale of pumping station should be determined by comprehensive analysis according to the project task, considering the near-term goal as the main objective, and considering the requirements of long-term development. At present, according to the *Design code for pumping station*, China divides pumping station projects into five levels, as shown in Table 7.1. Classification of industrial and urban water supply pumping stations shall be determined according to water supply object, water supply scale and importance. Pumping station grade determines the grade of its main buildings, secondary buildings and temporary buildings.

The grade of a pumping station project consisting of multi-stage or multi-pumping stations may be determined according to the grading index of the whole system. When the pumping stations are classified into different classes according to the classification index, the higher one shall prevail.

The pumping station buildings shall be classified according to the class to which they belong and their role and importance of the pumping station, and their levels shall be determined in accordance with Table 7.2.

7.2　Zoning and grading of irrigation and drainage district

7.2.1　Zoning of irrigation districts

The plane division of irrigation area should be determined according to the principle of the highest economic benefit. According to the experience of pumping station construction, there are several basic forms of plane division of irrigation district.

(1) Water lifting in one station and irrigation in one area. Only one pumping station is built in the whole irrigation area, and the whole irrigation area is controlled by one main channel. The pumping station heads the water to the elevation of the irrigation area and then supplies water from the channel system to the whole irrigation area. This scheme is suitable for small irrigation district with small area, low head, small ground elevation difference and short water delivery channel, as shown in Fig. 7.1. Most small irrigation district with small topographic elevation difference adopt this zoning.

(2) Multi station pumping and district irrigation. Because of the terrain or river partition in irrigation area, if the scheme of pumping irrigation with one pumping station is adopted, it is bound to increase the amount of channel and river crossing buildings, so as to increase the project investment. At this time, it is appropriate to disperse the station partition irrigation. As shown in Fig. 7.2.

(3) Multi station grading water lifting and zonal irrigation. In irrigation district with high elevation and large topographic change, in order to avoid the phenomenon that the water raised to the high level flows back to the low level for irrigation, resulting in energy waste, the whole area can be divided into several irrigation district with different elevations and the cascade pumping station is used for hierarchial pumping, each stage of irrigation district is controlled by a main channel, and the pumping volume of the former stage station must supply the pumping volume of the latter stage in addition to meeting the needs of the irrisluiced area. This method is suitable for district with large ground elevation difference or obvious terrain platform, as shown in Fig. 7.3. As the irrigation district itself is an integrated engineering system of water, machinery and electricity, it is necessary to apply the system engineering optimization method to determine the grading and partitioning problem of cascade pumping stations.

(4) Water lifting by stages and irrigation by zones in one station. For some irrigation district with large ground elevation difference but small area, one station of graded water heading and zoned irrigation can be used, i. e. several pumps with different head are installed at the same pumping station, and two outlet sumps with different elevation and lower elevation and corresponding channel water supply are provided.

7. 2. 2   Zoning and grading of drainage district

The zoning and grading of drainage district should satisfy the separation of internal and external water, high and low water as far as possible, and make full use of the conditions of gravity drainage.

Internal and external water separation is mainly due to the separation of flood and waterlogging to avoid floods from other rivers intruding into the polder; then the separation of inland river and farmland is required in the inland part of the drainage area. That is to say, when draining waterlogging, it is necessary to use the storage capacity of the external rivers, inland rivers and farmlands, but also to build gates, embankments, and hierarchical control.

Separation of high and low water is to require isoheight closure, high water discharge and low water discharge to avoid high water sumping to low land and prolonging drainage time, thus causing or aggravating waterlogging damage. At the same time, high water discharge is also to avoid high water sumping to low land and increase drainage head, which is conducive to reducing installed capacity and operation cost of drainage station. In addition, because high water often has the possibility of gravity drainage, it is necessary make full use of the opportunity to drain water by gravity drainage in time.

1. Zoning of drainage district

(1) Zoning of riverside lakeland and polder drainage area. Although the surface of lake area and polder area is flat, there will be a certain height difference, especially in large areas, where the conditions of gravity flow and efflux are different, so unified planning is required to carry out regional drainage. When zoning, proper consideration should be given to the original drainage system according to the topographic characteristics and the water level conditions in the drainage area. For district with high terrain and gravity drainage conditions, it is necessary to divide them into high drainage district by using gravity drainage conditions as much as possible; for district with low terrain and long-term high external water level above the field surface during drainage, it is divided into low drainage district, mainly pumping drainage; in the area between the two, the drainage mode combining gravity drainage and pumping drainage can be adopted.

(2) Zoning of semi-mountainous and semi-polder drainage district. This kind of area, after the hilly area or highland, front along the river, lake, commonly known as semi-mountainous and semi-polder district. Because the water level outside the flood season is higher than the farmland inside the embankment, and at the same time, the external

water at high altitude flows downstream, so it is easy to be flooded. When partitioning, closure ditches or flood diversion ditches should be excavated along the contour line of the design external water level elevation along the boundary of the mountain ridges. The mountainous and embankment are divided into high and low levels to reduce the installed capacity of the pumping station.

(3) Zoning of coastal and tidal river drainage district. For drainage systems in coastal and tidal reach district which are less affected by floods but more affected by tides, drainage district can be divided according to regional elevation. The district with the ground level higher than the average tidal level are smooth drainage district, the district with the ground level lower than the average low tidal level are non-smooth drainage district, and the area between the two is the semi-smooth drainage district. The smooth drainage area can drain by itself, while the non-smooth drainage area relies on pumping station for drainage, while the semi-smooth drainage area should consider increasing the drainage outlet, shortening the drainage path, and building tidal sluices at the outlet, using the intermittent spontaneous drainage at ebb tide. If such treatment still fails to meet the drainage requirements, consider building an efflux (waterlogging) station to close the sluice for drainage during the tide. In the planning of such district, the relationship between waterlogging, flood and tide level should be analyzed in detail to adjust the unreasonable drainage zones and outlets, expand the smooth drainage area and reduce the non-smooth drainage area as much as possible, and reduce the installed capacity and drainage costs of drainage stations.

2. Stage of drainage district

It includes three modes: one-stage drainage, two-stage drainage and combination of one-stage drainage and two-stage drainage.

(1) The one-stage drainage is to discharge the waterlogging water directly into the drainage area by the drainage station, as shown in Fig. 7. 4 (a); or to drain the waterlogging water into the waterlogging area first by the drainage station, and the waterlogging water in the waterlogging area will be discharged by itself when the external water level decreases, as shown in Fig. 7. 4 (b). The one-stage drainage method is suitable for drainage district with small drainage area, small power capacity and low head.

(2) Two-stage drainage mode, namely to build small stations in low-lying district to drain the waterlogging water into the waterlogging storage area, this type of station called two-stage stations or internal drainage stations, generally with low drainage head; while the waterlogging water in the waterlogging area needs to be drained by other stations, this kind of station called one-stage stations or external drainage stations, as shown in Fig. 7. 5. If the storage volume is large, besides heading and draining by pumping station, the water in the storage area can also be used to retain the water. After the external water level decreases, the sluice can be opened to drain the water in the storage area, and the

sluice station can be used to cooperate with the drainage to reduce the power capacity of the drainage station. Two-stage drainage is suitable for drainage district with large drainage area, complex terrain, uneven height and high head.

In the drainage area with large drainage pumping station, two-stage drainage mode should be preferred because of the large control range and complicated terrain in the drainage area. In this way, in layout, centralized and dispersed combination; in specifications, large, medium and small combination. In addition, the two-stage station (inland drainage station) has small capacity and flexible application, which can meet the needs of local lowland drainage or low-land drainage and high-land irrigation.

In the drainage area of coastal and tidal reach, if the tide recession drainage can be used and the flood storage volume is large, the inland drainage station can drain the low-land flood water first, and then open the sluice to drain the flood storage area after the tide recedes. In this case, the inland drainage station plays the role of the external drainage station, which belongs to the one-stage drainage mode. If the storage volume is small or the external water level is still relatively high after tide ebb, and can not rely entirely on gravity drainage by opening sluices, then an external drainage station should be set up to cooperate with pumping and drainage, which is the two-stage drainage mode.

In practical projects, the above two drainage modes are often not completely separated. Sometimes, the drainage station can discharge both the water in the flood storage area and directly. In operation, first drain the field and then drain the flood storage area. When the pumping station discharges the field, it is one-stage, while in the drainage and storage area, it is two-stage. See Fig. 7. 6.

The selection of drainage mode is an important work in the planning of drainage station, which should be determined by detailed technical and economic comparison. In addition, comprehensive utilization should be considered in the planning of drainage station. According to the need, a drainage station can be set up by reasonable arrangement of pumping station buildings. It is designed to not only pumping drainage and pumping irrigation, but also gravity drainage and gravity irrigation. It can be used for multiple purposes at one station to expand project benefits.

7. 2. 3   Site selection of pumping station

Site selection should be carried out simultaneously with the division of irrigation and drainage district. After zoning and grading is determined, the following factors should also be considered when selecting a specific site.

(1) The topographic site should be flat and open to facilitate the general layout of the pumping station buildings. The pumping station should be able to control the area of the whole district. The irrigation pumping station should be located in the upper reaches of the whole district with a higher terrain. The drainage pumping station should be located in the lower terrain in the drainage area with a shorter distance from the outlet of the drain-

age trunk river.

(2) The geological station shall be located on a solid foundation. Pump house and intake structure are usually built on deep excavation foundation surface. Foundations such as silt, quicksand, collapsible loess and expansive soil should be avoided to reduce the cost of foundation treatment.

(3) Rivers, lakes and reservoirs with abundant water and good water quality should be selected as water sources. For rivers, the straight reach and stable riverbed should be selected as far as possible. In case of bending river section, the concave bank with steep slope and sediment resistance should be selected, and the station construction in reach with shoals, tributaries inflow and bifurcation should be avoided. The drainage station should also select the lower water level section in the external river as far as possible in order to reduce the drainage head.

(4) The station site of the pumping station for comprehensive utilization of drainage and irrigation shall create conditions for combination of drainage and irrigation, and also provide sites for other purposes.

(5) The location of power station should be as close to the power grid as possible to reduce investment in transmission and substation projects.

(6) Traffic at other stations shall be convenient and close to townships and residential district to facilitate the transportation, construction and operation management of equipment and materials.

7.3   Design parameters of pumping station

The design parameters of pumping station are mainly the design discharge and head, which are the basis of pump selection and pumping station building design. Design parameters affect equipment investment, pumping station scale and benefits of the pumping station. Therefore, these parameters should be fully justified when they are determined.

7.3.1   Discharge of pumping station

1. Design discharge of irrigation pumping station

The design discharge of irrigation pumping station should be determined according to comprehensive analysis and calculation of crop irrigation system, irrigation modulus, irrigation area, water utilization coefficient of channel system and reserve volume in irrigation area under certain irrigation design criteria. Irrigation design standard is an index reflecting the guarantee degree of irrigation water source or pumping capacity for irrigation water in farmland, and is an important basis for determining the scale and design parameters of pumping station project. It should be carefully analyzed and determined according to water source condition, crop layout, hydrometeorological conditions, topography, soil quality, planning and economic conditions of local agricultural production, etc. The design standard of irrigation generally uses the guarantee rate of irrigation design or the number of drought resistance days as the design standard.

2. Design discharge of drainage pumping station

The design discharge of the drainage pumping station depends on the task of the pumping station. In plain polder district, drainage pumping station is mainly drainage, combined with irrigation, while drainage is mainly drainage, considering both drainage and groundwater level reduction. Therefore, the drainage flow should be taken as the design discharge of the drainage pumping station.

The drainage dischange is calculated according to certain drainage design criteria. If the design standard is too low, it will not play a significant role in mitigating floods, which will affect the production of agriculture, as well as the life of the people. If the design standard is too high, the project scale will be too large and the pumping station will not play a full role in normal years, resulting in waste. Therefore, the standard of drainage design should be determined by comprehensive analysis according to the development needs of agriculture and national economy, current situation of water conservancy facilities and human, material and financial conditions in the planning period.

Drainage modulus method, average exclusion method and storage algorithm are generally used to calculate the design discharge of drainage. Drainage days are mainly determined by the allowable flooding tolerance of crops. One-day rainstorm is usually used to exclude small area for 2 – 3 days and three-day rainstorm for large drainage area for 4 – 5 days. For dry crop district, drainage time should be shorter than that in paddy fields.

7.3.2 Head of pumping station

1. Head of irrigation pumping station

The upstream and downstream water levels of the pumping station include the water level of the outlet sump (upstream) and the water level of the inlet sump (downstream) . The upstream water level is determined by factors such as field elevation and irrigation flow in the irrigation area, and the downstream water level is determined by the water level of the source. When the elevation of channel headworks in irrigation area has been determined, the pumping station head can be calculated according to the characteristic water level of inlet and outlet sumps of the pumping station.

(1) Water level of intake sump. The water level of the intake sump can also be divided into the maximum water level, the design water level, the maximum operating water level, the minimum operating water level and the average water level. For pumping stations without or with short diversion channels, the method to determine the water level of the intake sump is shown in Table 7.3. If the diversion channel is long and the corresponding water level of the intake sump is determined, the hydraulic loss from the water source (intake) to the intake sump should be deducted. When fetching water from unstable rivers, the influence of riverbed changes on water level should also be considered.

1) The maximum operating water level at the inlet is used to check the working condition of pump and auxiliary power machine as well as the safety and stability of the project.

2) The inlet design water level is used to determine the pumping station design head.

3) The minimum operating water level at the inlet is used to determine the installation elevation of the pump and to check the safety of the project.

4) The maximum inlet water level is used for the pumping station which does not directly retain water in the pump house to determine the elevation of the floor of the machine room or the motor floor, design the water retaining wall and check the anti-floating and anti-sliding stability of the project. When the maximum water level occurs at the pumping station inlet, the pump unit does not necessarily need to be operated. Therefore, the standard for determining the maximum water level is not irrigation guarantee rate, but the level of the pumping station buildings.

5) The minimum water level at the inlet is determined by frequency analysis according to the data of dry water level of the source according to the standard required by the building grade of the pumping station. It is designed to check the safety of the project.

(2) Water level of outlet sump. The water level of the outlet sump of the irrigation pumping station is the head level of the irrigation channel calculated step by step from the tail of the channel to the head of the channel. Its determination method is as shown in Table 7.3.

(3) Characteristic head of irrigation pumping station. According to the above combination of water level of outlet pool and water level difference of intake pool, various head of pumping station can be calculated:

Design head＝Design outlet water level－Design inlet water level

Maximum head＝Maximum water level of outlet－Minimum operating water level of inlet

Minimum head＝Minimum water level of outlet－Maximum operating water level of inlet

The design flow requirement of the pumping station should be guaranteed under the design head. At average head, the pump should work in the BEP.

The average head can be calculated by Eq. (7.1) with weighted average head or by the average water level difference between the inlet and outlet sumps of the pumping station.

$$H = \frac{\sum H_i Q_i t_i}{\sum Q_i t_i} \qquad (7.1)$$

Where, H is the weighted average head, m; H_i is the difference of operating water level between the inlet and outlet pools in period i, m; Q_i is the heading water discharge in period i, m³/s; and t_i is the duration of period i (d).

2. Calculation of head of drainage pumping station

(1) Water level of inland and external rivers. The inland water level of the drainage

station generally includes the maximum water level of the inland river, the maximum operational water level of the inland river, the design water level of the inland river and the minimum operational water level of the inland river. The minimum operational water level and the maximum operational water level of the inland river are also called drainage level and stop-drainage level respectively.

The water level of the river outside the drainage station generally includes flood control level, maximum operating level of the river, design level of the river outside, minimum operating level of the river outside, etc. Its function in the design of the pumping station is similar to that of each water level of the irrigation pumping station. The methods for determining each water level are shown in Table 7. 4.

(2) Determination of characteristic head of drainage pumping station. When the water levels of inland and outside rivers are known, various drainage heads can be combined according to the opportunity of water level.

Design head＝Design external water level－Design internal water level

If there is a possibility that the maximum external water level will meet the minimum internal water level, the maximum head will be calculated as follows:

Maximum head＝Maximum operating external water level－Minimum
operating internal water level

If there is no possibility that the highest external water level and the lowest internal water level will meet, the following two water level combinations can be taken respectively, with the larger of them as the maximum head adopted:

Maximum head＝Maximum external water level－Design internal water level

or　　　　　Maximum head＝Design external water level－Minimum
operating internal water level

Minimum head＝Minimum operating internal water level－Maximum
operating external water level

7. 4　Selection of pumping station units

Pumps and matching power engines are the main equipment of the pumping station, which is called the host group. All auxiliary equipment and hydraulic structures are operation services of the main unit. Therefore, it is an important issue in planning whether the selection and matching of main unit is reasonable, which directly affects the benefit of pumping station, investment in civil engineering and electromechanical equipment, equipment utilization rate, etc.

7. 4. 1　Pump Selection

1. Type selection requirements

The pump type selection shall meet the following requirements: can meet the design discharge and design head requirements of water supply and drainage and irrigation under certain design standards; the pump operates in the BEP; the size and number of units shall make the pumping station less expensive, convenient for maintenance and management, and less operating costs; the pump runs stably and safely with good cavitation performance.

2. Comparison and selection of pump types

The main pump model of the pumping station is mainly selected according to the head and discharge, and the supply of existing pump products is also considered. For high-head pumping stations, double-suction centrifugal pumps or multi-stage centrifugal pumps are basically selected, and the capacity of the single unit is controlled by the number of units. For medium and low head pumping stations, there are sometimes more pump types to choose, so when selecting the type, technical and economic comparison should be made on the structure type, installation method, etc.

Selection charts of various pump performance specifications include pump samples and comprehensive spectrograms of pump series, etc., and catalogues of products are also available for reference in various pump plants.

3. Steps of pump selection

(1) According to the planning requirements of the water supply (irrigation) and drainage district, determine the two parameters of discharge and head of the pumping station, which are generally determined by the water supply (irrigation) and drainage planning. According to the water conservancy planning, the key parameters such as the upstream and downstream water levels and the total design discharge of the pumping station are determined.

(2) Number and model of primary pump.

1) To determine the number of pumps, the size of pumps and the investment of civil engineering, we need to attach great importance to it. Too few sets, too low assurance rate; too many sets, high operating costs, inconvenient management. From experience, it is generally appropriate to use 4－8 sets, preferably no more than 8－10 sets. After determining the number of units, the design discharge of single pump can be calculated at $Q_{station}/n_{units}$.

2) Net head calculation of pumping station calculates net head of pumping station by planning given upstream and downstream water levels, including design net head, maximum net head, minimum net head and other characteristic heads.

3) Estimation of pipeline loss head since the pump model has not been selected at this time, the inlet and outlet pipelines of the pumping station have not been designed and arranged, and the pipeline loss head is estimated when the pump is selected initially. The pumps of high-head pumping station adopt the method of selecting pumps with equal discharge and increased head, that is, the estimated pump head is the sum of the net head

and estimated loss head under the design discharge. The loss head of pipeline should be 20%–30% of the net head. Engineering practice, especially the pump selection experience of the first stage pumping station of the Eastern Route of South-to-North Water Diversion Project, shows that the optimum operating point of the low-head pumping device deviates greatly from the small discharge, and the corresponding high-efficiency device head and pump head are close, while the pump discharge can be increased appropriately, about 10% can be taken. The selection method of low-head pump with equal head and larger discharge is more accurate, which can avoid various operation problems caused by long-term low-head and large-discharge operation in actual operation.

4) Preliminary selection of pump type is based on the design discharge of a single pump and the design head of the pump. More than two pump types are selected from the performance table, the comprehensive performance curve of the pump, the pump type spectrum or from the computer database. According to the pump design head and discharge, determine the pump speed or operation angle (see Fig. 7. 7).

(3) Power engine matching shall select power matching power according to the most disadvantageous condition of pumping station, and determine the model of power engine.

(4) Design and layout of inlet and outlet pipelines of pumping station.

1) Pipeline layout. According to the size of pump type selected by pump sample type, arrange the inlet and outlet pipes, and select appropriate pipe accessories and cut-off facilities.

2) Pumping station pipeline characteristics calculate local and along-the-way losses of pipelines respectively, then calculate resistance coefficient S of pumping station inlet and outlet pipelines, and draw pipeline characteristic curve.

3) Required pump characteristics of pumping station. The difference of water level between upstream and downstream under each operating condition (net pump H_{st} of pumping station) plus pipeline loss (SQ^2) are added to draw the required pump curve under each characteristic net pump.

(5) Check of working point.

1) The performance curve ($H-Q$) of the primary pump type and the curve H_r-Q of the required lift are drawn in the same coordinate system. The intersection of the two curves is the working point D, E and F of the pump (see Fig. 7. 8)

2) Check whether the design net pump operating point (point D in Fig. 7. 8) is in the selected pump type high-efficiency area, and the discharge Q_D of the design operating point is larger than the planned design discharge $Q_{\text{single pump}} = Q_{\text{pumping station}}$/unit number.

3) Check whether the power of the power engine is overloaded under the most unfavorable condition (centrifugal pump is the minimum pump condition, point F in Fig. 7. 8).

If one of the above conditions does not meet the requirements, the pump shall be reselected or the layout of the inlet and outlet pipelines shall be changed. The performance parameters of pump type selection schemes of the two pumping stations focus on comparing the efficiency and the $NPSH_r$ at the design duty point.

4. Some problems in pump selection

(1) Number of units. Generally, two pumps can be selected when the discharge of pumping station is less than $1m^3/s$, and three to eight pumps can be considered when the discharge of pumping station is greater than $1m^3/s$. When the water supply assurance rate is high, standby pumps should be considered, but the total number should not exceed 10 sets.

The number of reserve units shall be determined according to the importance of water supply and annual utilization hours, and shall meet the normal maintenance requirements of units. For important urban water supply pumping stations, when there are less than 5 working units, one standby unit shall be added, and when there are more than 5 working units, another standby unit shall be added; for agricultural pumping stations, standby units may be appropriately reduced, and relevant specifications may also be referred to.

(2) Installation form of pump. There are three types of pump installation: vertical, inclined and horizontal. Horizontal pumps generally have installation elevation above the inlet water surface, small excavation, lower installation requirements than vertical pumps, convenient maintenance and good working conditions. However, the horizontal unit covers a large area and usually needs to be vacuum pumped before starting. Vertical pump covers a small area, the impeller is submerged below the water surface, there is no water intake pipeline or short water intake pipeline, and it is convenient to start. However, the installation requirements are high, the pump house height is large, and the project cost is high. The inclined pump is between the vertical and horizontal pumps.

(3) Select pump with good anti-cavitation performance. Selecting a pump with good anti-cavitation performance can improve the installation elevation of the pump, reduce the excavation depth of the pumping station and save the project investment.

(4) Comprehensive consideration of multiple factors. The selection of pumps has a direct relationship with the civil structure of pumping stations, which often needs to be considered together with the civil engineering design scheme and decided after comprehensive comparison.

7.4.2　Selection of primary movers

After the model of the pump is determined, it is necessary to select a suitable power and rotating device for the pump. The most commonly used driving pump is motor, followed by diesel engine. When purchasing a pump, the pump manufacturer will usually supply the matching products, and the user should generally carry out calculation review or self-selection.

1. Required power

When determining the matching power of the power engine, it is calculated according to the maximum shaft power under the most unfavorable working conditions within the working range of the pump. The matching power is calculated according to the following formula:

$$N_{mt} = K \frac{\rho g Q H}{1000 \eta_p \eta_{dr}} \quad (\text{kW}) \tag{7.2}$$

Where, Q is the discharge corresponding to the maximum shaft power (m^3/s) under the most unfavorable operating condition of the pump; H is the head (m) corresponding to the maximum shaft power under the most unfavorable operating condition of the pump; η_p is the efficiency corresponding to the maximum shaft power under the most unfavorable operating condition of the pump; η_{dr} is transmission efficiency (0. 9 – 1. 0) . K is spare coefficient of power machine, selected according to Table 7. 5.

The spare coefficient values in Table 7. 5 can be selected according to the principle of small pump taking large value and large pump taking small value, and the determination of matching power should also conform to the rated capacity of the power machine.

2. Some problems of motor type selection and matching

Motor is one of the main equipment of electric pumping station, and its correct selection will directly affect the efficiency of pumping device and the safe operation of pumping station. Motor selection mainly considers the following aspects:

(1) Selection of motor type. The commonly used motors in pumping stations are asynchronous motor and synchronous motor. The squirrel cage type asynchronous motor has the advantages of simple structure, reliable operation, convenient maintenance, low price, and easy to realize automatic control, so it is used more. Although the starting current is large, the instantaneous voltage drop is large, and the starting torque is small, because the pump belongs to light load starting, the use of such motors has met the engineering requirements. It is necessary to equip starting equipment for step-down starting of large capacity rat cage asynchronous motor. When the power grid capacity cannot meet the starting requirements of squirrel cage motor, winding motor can be considered.

Although the synchronous motor has high cost, complex structure and troublesome maintenance and management, it has high efficiency and power factor, and can also be used for phase adjustment operation to transmit reactive power to the power grid. At the same time, it has the characteristics of large starting moment and strong adaptability to the fluctuation of network voltage, which is suitable for large pumping stations and occasions with long service time.

(2) Installation form of motor. The installation form of the motor is generally consistent with that of the pump, that is, the horizontal pump is equipped with a horizontal motor, and the vertical pump is equipped with a vertical motor. In engineering, if horizontal motor is also used to equip vertical pump, indirect transmission mechanism is needed.

(3) Selection of motor voltage. Select the power supply voltage according to the

power of the motor: 0. 4kV three-phase alternating current is selected for the power below 200kW; 10kV (or 6kV) three-phase alternating current is selected for the power above 300kW; two voltages are selected for the power between 200 and 300kW, which can be reasonably determined by combining the local power grid conditions and technical and economic comparison.

(4) Selection of motor speed. The speed of the motor matched with the pump must meet the requirements of the speed of the pump. If the difference between the speed of the pump and the rated speed of the motor sample is less than 2%, the direct connection mode can be used directly, otherwise the indirect transmission device should be used.

7.5　Junction layout of pumping station

Junction layout of pumping station is the general layout of all the buildings of the pumping station, including: intake, diversion, suction buildings; pump house; transformer substation, outlet buildings and other positions and layout forms between each other. The layout of pumping station junction has a great influence on the cost of the whole project and the operation and management of the pumping station. The type of layout to be adopted depends on the requirements of the drainage and irrigation tasks, topography, water system conditions, and the original project conditions, and the pump unit to be selected comprehensive consideration of the performance and structural characteristics, etc. , in order to obtain an economical and reasonable plan.

The intake structures of the pumping station include the forebay, intake sump and intake pipe. For irrigation pumping stations with diversion channels, diversion channels are also water intake structures; for pumping stations built on sediment-laden rivers, sediment sinks are often built in the forebay part. Water outlet structures include water outlet pipelines (passage), water outlet pools, etc. Pumping station house (pump house) is a building which installs pump, power machine and auxiliary equipment as well as auxiliary equipment of pumping station, and is the main project in each building of pumping station.

7.5.1　Some problems of junction layout

The main factors affecting the layout of the junction are the purpose of the station construction, the type of water source, the variation of water level and water quality, as well as the topographic, geological and hydrological conditions of the station location.

(1) The layout of the pivots must comply with the requirements of the regional governance planning, and compare the schemes according to the site conditions and referring to the experience of the established pumping stations.

(2) To create good water flow conditions, the flow of diversion and inflow should be stable, avoiding the backflow, dead water zone and flow whirlpool.

(3) To give full play to the role of each building, that is, to maximize the benefits with the fewest buildings as possible.

(4) The selection and correlation of the types of each building shall ensure safe operation and convenient management, and leave room for the expansion of the pumping station.

(5) Minimize excavation, occupation of farmland and demolition of original buildings.

(6) Consider comprehensive utilization as far as possible, such as combination of drainage and irrigation, combination of pump drainage and gravity drainage, combination of pumping station and traffic, etc.

7. 5. 2   Arrangement type of irrigation pumping station junction

1. Layout of pumping stations with diversion structures

Fig. 7. 9 shows the layout of pumping station junction with diversion open channel, which is suitable for taking water from rivers and lakes, siting with slow slope and small variation of water level.

2. Layout of pumping station without diversion structures

When the shore of the water source is steep, the water level changes little, or the irrigation area is close to the water source, the pumping station and the water intake building are often combined and built directly on the shore of the water source or in the water source (Fig. 7. 10) . Therefore, the pump house itself needs to retain water. The advantages of this form are that the diversion structures are cancelled and the maintenance workload is reduced. The disadvantage is that the structure of pump house is complex and flood control safety needs to be considered.

7. 5. 3   Junction layout type of drainage pumping station

The layout pattern of pumping station junction for simple drainage is usually shown in Fig. 7. 11. The outlet of the outlet pool is close to the river embankment, and the outlet of the outlet pool is connected with the drainage culvert. The waterlogging pumped to the outlet pool by the pump is discharged into the external river through the drainage culvert. There are also open channels for discharging water without drainage culverts. Discharge sluice can play a flood control role. When the water level of an external river is high and there is no need for pumping and drainage, the discharge sluice is closed.

According to the relationship between pump house and drainage sluice, it can be divided into two types of layout, i. e. split-construction type and combined-construction type.

(1) Separate pumping house and drainage sluice are constructed separately, and the diversion channel draws water from the drainage river. This form is applicable to the original drainage sluice, but the problem of waterlogging cannot be solved by gravity drainage alone, so it is necessary to build a pumping station to drain water during closing the sluice.

(2) The combined pumping house is built together with the drainage sluice. When the water level of the inland river is higher than that of the external river, the sluice is opened for drainage, and when the water level of the external river is higher than that of the inland river, the sluice is turned off for drainage.

7.5.4 Junction layout type of drainage and irrigation combined pumping station

1. Layout of one station and four sluice stations

For pumping station projects with both irrigation and drainage tasks, there are many supporting buildings, and more attention should be paid to rational layout. In order to arrange reasonably, the necessary supporting buildings and their regulation and application relations should be determined first. The layout of the pumping station junction shown in Fig. 7.12 (a) is called "one station and four sluices", which consists of four control sluices and one pumping station. This arrangement can meet the needs of pumping irrigation, pumping drainage, gravity guiding and gravity discharging.

When pumping water for irrigation, open sluice 1 and sluice 3, close sluice 2 and sluice 4.

When pumping water and draining water, open sluice 2 and sluice 4, close sluice 1 and sluice 3.

During gravity drainage and diversion, sluice 3 and sluice 4 are opened.

Open sluice 1 and sluice 2 when gravity draining.

The advantages of the layout form of one station and four sluices are good water intake conditions, convenient drainage and irrigation dispatch, and less contradictions. Its disadvantage is that there are more hydraulic structures and higher investment in civil engineering.

2. Combination type of bidirectional flow sluice station

There are many types of pivot layout of drainage and irrigation combined pumping station, which are often closely related to the direction of ditches. In the construction of large pumping stations, in order to meet the needs of gravity diversion, gravity drainage and two-way pumping, the flow channel is often designed as an X shape, that is, the combination type of two-way flow sluice station, as shown in Fig. 7.13. The lower layer of the station building is not only an inflow channel, but also a culvert for diversion or drainage. The upper layer is an outflow channel, and both the inflow and outflow are bidirectional. If the water level in the external river is low, sluices 2 and 4 can be opened, and the water in the inland river can flow into the external river spontaneously; if the water level in the external river is high, sluices 1 and 4 can be closed, sluices 2 and 3 can be opened and discharged into the external river. When irrigating, if the water level of the external river is high, open sluices 2 and 4 to irrisluice the water from the external river; if the water level of the external river is low, close sluices 2 and 3, open sluices 1 and 4 for pumping irrigation. The station house of this kind of pumping station directly retains water and is a dike-body pumping station, which is suitable for occasions with low pump and small variation of internal and external water levels.

Compared with the layout form of "one station and four sluices", the combined layout of two-way flow sluice station covers a small area, saves project investment and facilitates centralized management. It has the advantages of simple structure, bidirectional pumping, combination of pumping and drainage, and high operation efficiency, so it has been widely used.

第8章 进水建筑物

泵站进水建筑物主要包括取水头部、进水涵闸、引水明渠或暗管、前池或集水井、进水池或吸水室等。大中型泵站还包括进水流道。进水建筑物布置和水力设计合理与否，直接影响水泵的工作性能和泵站效率。为保证泵站的安全经济运行，泵站进水建筑物应满足以下要求：①足够的进水能力；②良好的进水流态；③一定的稳定性和强度；④节省土建投资。

8.1 取水与引水建筑物

8.1.1 取水建筑物

采用涵管从水源中取水的建筑物称为取水头部。其结构形式较多，包括重力式、沉井式、框架式、悬臂式、底槽式及隧洞式等。

从水源岸边取水的建筑物有进水涵、闸，开敞式取水口等。在多泥沙河流中取水，应选择有利位置，取用含沙量少的表层水，并采用导流设施将含沙量大的底层水导走，同时在引渠适当位置设置沉沙池。

8.1.2 引水建筑物

当泵站建于水源附近或排水区岸边确有困难时，应设置引水建筑物，可采取引水明渠（引渠）或利用现有河道、压力涵管等形式。其中，泵站引渠作为连通水源（或排水区）与泵房的明渠，其主要作用如下：

(1) 使水流平顺地进入前池。

(2) 避免泵房与水源直接接触，简化泵房结构，便于施工。

(3) 多泥沙的水源取水泵站，为沉沙提供条件。

对引渠的要求是：

(1) 能随时输送泵站所需的流量。

(2) 能适应泵站抽水流量的变化。

(3) 水力损失要小。

1. 引渠路线的确定

引渠路线的选择应根据选定的取水口及泵房位置，结合地形地质条件、施工条件及挖填方平衡等多方面的因素，经技术经济比较后确定。渠线应避开地质构造复杂、渗透性强和有崩塌可能的地段，渠身宜坐落在挖方地基上，少占耕地；为了减少工程量，渠线宜顺直，如需设弯道时，土渠弯道半径不宜小于渠道水面宽度的5倍，石渠及衬砌渠道弯道半径不宜小于渠道水面宽度的3倍，弯道终点与前池进口之间宜有直线段，长度不宜小于渠道水面宽度的8倍。

2. 引渠的类型

(1) 自动调节引渠。渠道顶部尾部高程基本相同，渠顶线高于水源最高水位线，因此渠道内无论通过多少流量，其水位都不会超过堤顶而发生漫溢现象，所以引渠无须设置控制建筑物。其优点是渠道有一定容积，可以减缓水泵启动与停机时渠道内发生较大幅度的水位涨落；其缺点是挖方量大，泵房应有防洪措施。

(2) 无自动调节引渠。引渠堤顶线平行于渠道底坡线，当渠中通过设计流量时，渠内水流为均匀流，水面线平行渠底，水深等于正常水深。当抽水流量小于设计流量时，渠中水流处于非均匀流状态，水面曲线为壅水曲线，为保证引渠不发生漫溢，常在引渠末端适当位置设置溢水建筑物，或在渠首设控制闸。其优点是挖方少，适用于长距离引渠；泵房前无须设置防洪设施；其缺点是需增加控制水位建筑物。

3. 引渠断面的设计

引渠的设计方法与一般输水渠道基本相同，即按均匀流设计，按不冲不淤条件校核。

4. 水泵启动、停机对引渠内流动的影响

在水泵启动时，渠内水面会出现跌落；而在水泵停机时，渠内水面则将产生壅高。水面跌落和壅高均以波的形式出现，跌落时产生逆落波，壅高时产生逆涨波（图 8.1）。在确定水泵吸水管进口要求的淹没深度时，应考虑最后一台机组启动所产生的水面降落（即逆落波）的影响；在确定渠道顶部高程时，则应考虑机组突然停机所产生的水面壅高（即逆涨波）的影响。

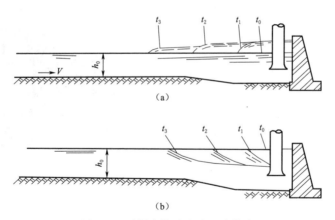

图 8.1 引渠中的逆涨波和逆落波

Fig. 8.1 Positive surge and negative surge in diversion channel

（a）逆涨波；（b）逆落波

(a) Reverse rising wave；(b) Reverse falling wave

逆落波高度可按下式近似计算

$$\Delta h_{落} = 2 \frac{Q_1 - Q_0}{B \sqrt{gh_0}} \quad (m) \tag{8.1}$$

式中：Q_1、Q_0 分别为水泵机组启动前、后渠内的流量，m^3/s；B 为渠中平均水面宽度，m；h_0 为水泵机组启动或停机前的渠中水深，m。

逆涨波高度可按下式近似计算

$$\Delta h_{涨} = \frac{(v_0 - v')\sqrt{h_0}}{2.76} - 0.01h_0 \ (\text{m}) \tag{8.2}$$

式中：v_0 为突然停机前引渠末段流速，m/s；v' 为突然停机后引渠末段流速，m/s。

水面壅高产生的逆涨波可能影响引渠堤顶和泵房的防洪高程，水面跌落产生的逆落波可能减小水泵进水口的淹没深度，导致水泵发生汽蚀和机组振动，甚至影响水泵正常运行。

综上所述，在引渠设计时要充分反映泵站工作的特点，即要求引渠能为水泵安全运行创造良好的水流条件。引渠设计流量的确定，应尽可能使引渠处于等速流或稍有壅水现象，在任何情况下都不宜有过大的水面降落。渠道的长度和纵坡亦应通过工程造价及运行费用综合计算进行综合选择。

8.2 前 池

引渠与进水池之间的衔接段称为前池，它的作用是：将引渠水流平顺均匀地输送给进水池，为水泵提供良好的吸水条件；同时具有一定容积，减少水泵启动和停机时水位的变化。图8.2所示为典型多机组泵站前池与引渠及进水池的连接情况。

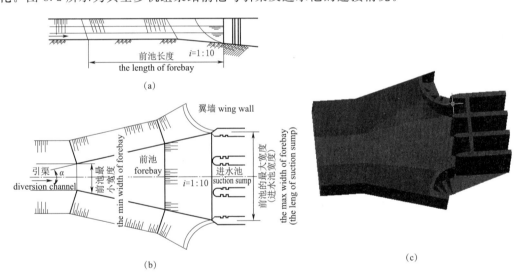

图 8.2　前池与引渠及进水池的连接

Fig. 8.2　Connection between forebay, diversion channel and suction sump

(a) 剖面图；(b) 平面图；(c) 透视图

(a) Section；(b) Plan；(c) Perspective

8.2.1　前池的作用与类型

根据前池与进水池的位置，分为正向进水前池和侧向进水前池两种。

正向进水前池中心线与引渠、进水池的中心线重合，水流方向也基本一致［图8.3（a）］。侧向进水前池的水流与进水池水流方向正交或斜交［图8.3（b）］。

在正向和侧向进水前池中，又可以分为有隔墩（图8.4）和无隔墩两种类型。其中隔

墩的作用为:

(1) 分流导水,使水流平稳,水量均匀地进入每个进水池,特别在不对称开机时避免前池水流大面积晃动、回流及漩涡。

(2) 如果前池扩散角过大,可以减小水流的实际扩散角,并缩短池长。

(3) 隔墩还可以作为闸墩和工作桥墩用。

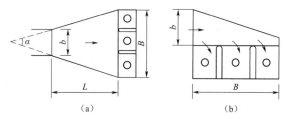

图 8.3　前池的两种基本类型

Fig. 8.3　Two basic types of forebay

(a) 正向进水;(b) 侧向进水

(a) Front inlet;(b) Side inlet

8.2.2　前池中的不良流态与水力设计要求

1. 前池不良流态

(1) 正向进水。设计合理的正向进水前池,当水泵全部或部分对称开机时,流态较好,所有泵站应尽可能采取正向进水前池。但当部分水泵不对称运行时,或当前池水位较低、泵站流量较大、前池扩散角过大、引渠来流的流态不良时,正向进水前池也会出现以下几种不良流态:

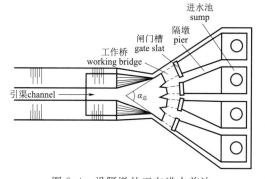

图 8.4　设隔墩的正向进水前池

Fig. 8.4　Front forebay with piers

1) 主流偏斜,一侧出现回流及死水区,部分进水池进水成为侧向进水。

2) 主流居中,前池两侧产生脱壁回流,中间进水池受主流顶冲,两边进水池成为侧向进水池,如图 8.5 (a) 所示。

3) 从前池进口开始发生扩大的折冲水流,池内流态十分紊乱。

第 1、2 种流态,在前池扩散角较大、水位较低而泵站流量较大时容易发生。第 3 种流态为引渠来流具有折冲状态所引起。

(2) 侧向进水。侧向进水前池中的水流通常为不良流态。由于水流在前池内要改变流向,易产生边壁脱流、大面积回流,在泵站进水池前产生不均匀行近流速分布,进而导致泵站各机组进水池隔墩头部背水面产生漩涡,池内流向偏向一边,水泵吸水口附近产生水面涡等,如图 8.5 (b) 所示。

2. 前池水力设计要求

试验研究表明,前池内的不良流态将严重影响进水池内的流态,导致水泵能量性能和汽蚀性能下降,同时大面积回流还将引起前池内的局部淤积,而泥沙淤积又会进一步加剧不良流态的发展。前池的水力设计要求要保证水流顺畅、扩散平缓,无脱壁、回流或漩涡现象,同时还要考虑尽可能节省土建投资。

8.2.3　前池尺寸的确定

1. 前池扩散角 α

前池扩散角 α 不仅影响前池流态,而且影响工程量。α 过大,则前池池长短、工程量小,但水流来不及扩散,易导致脱壁回流;α 过小,则前池水流扩散平缓,但池长要增加。

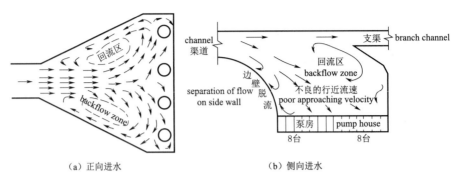

（a）正向进水　　　　　　　　　（b）侧向进水

图 8.5　前池内的不良流态

Fig. 8.5　Poor flow pattern in the forebay

（a）正向进水；（b）侧向进水

（a）Front inlet forebay；（b）Side inlet forebay

水流在渐变段扩散流动时有一个天然扩散角，即不发生脱壁回流的临界扩散角，其值可根据以下半经验半理论公式进行计算

$$\tan \frac{\alpha}{2}=0.065 \frac{1}{\sqrt{Fr}}+0.107 \tag{8.3}$$

其中

$$Fr=\frac{v}{\sqrt{gh}}$$

当 $Fr=1$ 时，即水流处于急流与缓流之间的临界状态时，$\alpha=20°$。由于前池中的流动通常为缓流，$Fr<1$，故其扩散角 α 可以大于 $20°$。根据实际工程经验，前池扩散角的取值一般为 $\alpha=20°\sim40°$。

2. 前池池长 L

前池池长可由引渠末端底宽 b、进水池总宽 B 及选定的前池扩散角 α 算得，即

$$L=\frac{B-b}{2\tan \frac{\alpha}{2}}\ (\text{m}) \tag{8.4}$$

3. 前池底坡 i

由于水泵淹没深度的要求，进水池池底的高程一般低于引渠末端的渠底高程，因此，前池池底为斜坡，使其在立面方向上起连接作用，其坡度为

$$i=\frac{\Delta H}{L} \tag{8.5}$$

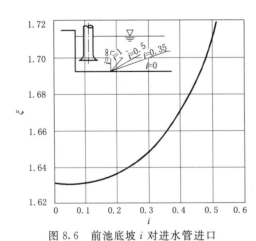

图 8.6　前池底坡 i 对进水管进口
阻力系数 ξ 的影响

Fig. 8.6　Effect of forebay bottom slope i
on inlet drag coefficient of inlet pipe

式中：ΔH 为引渠末端渠底高程与进水池池底高程之差。

根据有关试验资料，前池底坡 i 对进水流态有一定影响，图 8.6 所示为进水管进口阻力系数 ξ 与前池底坡 i 的关系。由图可知，前

池底坡 i 越大，阻力系数 ξ 越大。当 $i<0.3$ 时，阻力系数 ξ 的变幅较小；当 $i>0.3$ 时，阻力系数 ξ 的变幅较大。因此 i 应在 $0.2\sim0.3$ 范围内选取。当 $i<0.3$ 时，为节省工程量，可将前池的前段做成水平，靠近进水池的后段做成斜坡。

8.2.4 前池流态的改善

正向进水的前池内流态较好，通常不必采取整流措施，只有当前池扩散角过大或经常不对称开机时，需设置导流墩及底坎、立柱等来改善前池内的不良流态。

侧向进水的前池内流态较差，一般均需采取适当的整流措施。图 8.7 所示是某泵站侧向进水前池中采用立柱和底坎整流措施改善进水流态的情况。试验和现场运行表明，适当位置采用立柱及底坎整流后，破坏了原方案的大面积回流区，改善了 2 座泵站进水池前行近流速分布，实现了大部分进水池正向均匀入流，从整体效果看，水泵的运行条件有了很大改善。

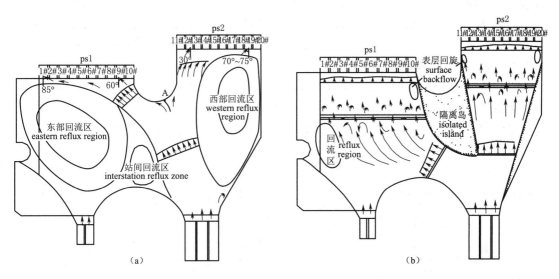

图 8.7 某泵站侧向进水流态改善实例

Fig. 8.7 An example of side inlet flow improvement at a pumping station

（a）无整流措施；（b）设立柱和底坎

（a）Non-rectifying measure；（b）Set up the column and sill

8.3 进 水 池

8.3.1 进水池的作用与设计要求

开敞式进水池是供水泵或进水管直接吸水的构筑物，其主要作用有：

（1）进一步调整从前池进入的水流，为水泵进口提供良好的进水条件。

（2）在检修水泵或进水管时截断水流。

（3）拦截来流中的污物。

水泵叶轮是在进口水流为均匀来流的假定下进行设计的。对于立式轴流泵及导叶式混流泵，进水池作为湿室型泵房的下层，也称为泵室。由于水泵叶轮室紧靠吸水喇叭口，吸水喇叭口导向作用较差，进水池内流态对水泵工作性能影响较大。对于带进水管的卧式水泵，由于进水管可以自动调整不均匀的进水流态，故对进水池中流态要求比立式泵要求稍低，但是，进水池中的有害漩涡仍会对水泵稳定安全运行产生影响。

因此，对进水池的设计要求是：

（1）池内流态良好，无明显漩涡回流。

（2）满足泵站进水要求。

（3）便于清淤和管理维护。

8.3.2　进水池的主要几何参数

矩形进水池的主要几何参数包括：进水池宽度 B、喇叭口悬空高度 C、后壁距 T、池长 L 及喇叭口淹没深度 h_s，如图 8.8 所示。图中，D_L 为喇叭口进口的直径，目前大多以 D_L 为基本参数表示进水池的各几何参数。

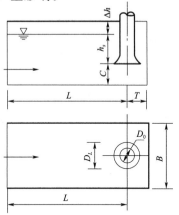

图 8.8　进水池的几何参数
Fig. 8.8　Geometric parameters of suction sump

8.3.3　进水池内基本流态

水流从四面汇集进入喇叭口是进水池流动的基本特征。如图 8.9 所示，在水泵运行时，一部分水流从正面进入喇叭口，一部分水流从两侧进入，还有一部分则绕过两侧，从后面进入喇叭管。

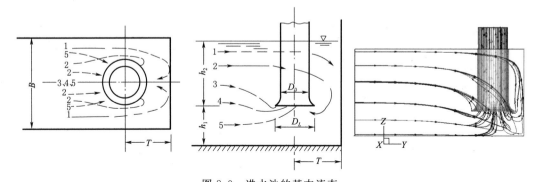

图 8.9　进水池的基本流态
Fig. 8.9　Basic flow pattern of suction sump

8.3.4　进水池内不良流态及漩涡

进水池内的不良流态主要是各断面流速分布不均匀及出现各种形式的漩涡，其中对水泵运行危害最大的是漩涡进入水泵内。

1. 进水池中漩涡类型

在进水池设计不当或水泵吸水管淹没深度不够的情况下，进水池内可能产生漩涡。

从漩涡发生的位置加以区分，可将漩涡分为附底涡、附壁涡和水面涡 3 种类型，如图 8.10 所示。

根据水面涡的吸气情况，通常又将其分为四种型式，如图 8.11 所示。

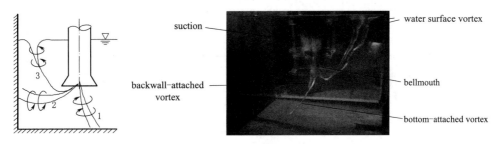

图 8.10　漩涡的类型

Fig. 8.10　Types of vortices

1—附底涡 bottom-attached vortex；2—附壁涡 wall-attached vortex；3—水面涡 water surface vortex

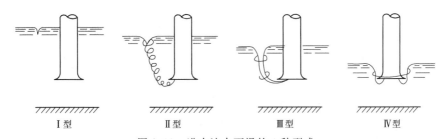

图 8.11　进水池水面涡的 4 种型式

Fig. 8.11　Four types of surface vortex in suction sump

Ⅰ型涡：水流旋转速度较慢，水面仅形成较浅的漏斗，尚未将空气带入水泵。

Ⅱ型涡：水流旋转速度较快，水面漏斗较深，已将空气断续地带入水泵，对水泵性能仅有轻微影响。

Ⅲ型涡：水流旋转速度很快，水面漏斗已伸入吸水管，空气连续进入水泵，对水泵的运行产生严重影响。

Ⅳ型涡：水流在吸水管周围急剧旋转，漩涡中心与吸水管中心一致，大量空气进入水泵，水泵机组产生剧烈振动以至无法运行。

当喇叭口淹没深度 h_s 较大时，只可能发生Ⅰ、Ⅱ型涡。当 h_s 减少时，出现漏斗状Ⅲ型涡，当 h_s 进一步减少时，Ⅳ型涡产生。Ⅰ～Ⅲ型涡称为局部涡，Ⅳ型涡称为柱状涡或同心涡。

2. 漩涡防治措施

进水池内的漩涡不同程度影响水泵的运行，为此，除了在前池内采取设置立柱或底坎等整流措施外，还可以通过在进水池内设置隔板、导流锥等措施防止漩涡的发生（图8.12）。

8.3.5　进水池尺寸的确定

进水池几何尺寸的确定目前大多依赖于试验结果，由于试验条件的差异，所得试验数据常常并不一致。随着计算流体动力学（CFD）的迅速发展，国内已开始采用理论计算的方法研究解决进水池的水力设计问题。

1. 进水池宽度 B

进水池宽度过小，会使池中流速加快、水头损失增加，增大了水流向喇叭口水平方向

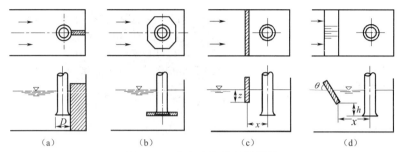

图 8.12　进水池各种防涡措施

Fig. 8.12　Anti-vortex measures in suction sump

（a）管后垂直隔板；（b）管口水平隔板；（c）管前垂直隔板；（d）管前倾斜隔板

（a）Vertical baffle behind tube；（b）Horizontal baffle at nozzle；

（c）Vertical baffle in front of tube；（d）Tilting baffle in front of pipe

收敛时的流线曲率，易诱发漩涡。进水池宽度过大，不仅会增加土建投资，而且降低了水池的导向作用，易在池中形成偏流、回流而产生漩涡。根据对进水池基本流态的分析，水流是从四周进入喇叭管的，若池宽过小，势必影响一部分水流顺利地从喇叭管两侧及后部进泵，故而需要一定的池宽，但池宽也不要过大，否则会增加土建投资。

池宽的确定，除需考虑水力条件外，还要考虑机组安装、维修的要求，一般要求 $B = (2\sim3)D_L$。在一池多泵的情况下，为减少水泵之间的相互影响，相邻两台水泵之间的距离可适当加大，取 $(3\sim4)D_L$。

2. 喇叭口悬空高度 C

喇叭口悬空高指吸水喇叭管进口至进水池底部的距离，其取值对喇叭管附近流态和土建投资的影响都非常显著。进水池的水流是从四周进入喇叭口的，合适的悬空高度对于形成这样的流动并使水流基本均匀地进入喇叭口至关重要。悬空高度过大，不仅增加了挖深和投资，而且有可能形成喇叭管的单面进水，导致水泵进口的流速和压力分布不匀，降低水泵的能量性能和汽蚀性能，有时甚至还会产生附底涡或附壁涡。悬空高度过小，则压缩了喇叭口下方的圆柱面，导致流入喇叭口的水流流线过于弯曲，喇叭管进口的水力损失急剧增加，也会使水泵进口的流速、压力分布不匀，增加漩涡发生的可能。悬空高度对进水流态的影响如图 8.13 所示。

综合国内外相关研究成果，建议悬空高度 $C = (0.6\sim0.8)D_L$。

3. 后壁距 T

后壁距指吸水管中心至进水池后壁的距离。通过分析进水池的基本流态可以注意到，有一部分水流是从喇叭管的后部进入喇叭管的，因此，必须留有一定的后壁空间。过小的后壁距必将导致不均匀的流态和较大的喇叭管进水损失。过大的后壁距不仅是不必要的，而且还增加了水流在后壁空间的自由度，从而加大了吸气漩涡产生的可能。研究结果表明，后壁距越大，所需的淹没水深越大。

合理的后壁距还应满足喇叭管安装的要求。建议后壁距 $T = (0.8\sim1.0)D_L$。

4. 进水池长度 L

进水池长度指从进水池进口至吸水管中心的距离。进水池长度应能调整池中流速使之

均匀，并保持一定的容积避免水位急剧跌落。一般按下式确定

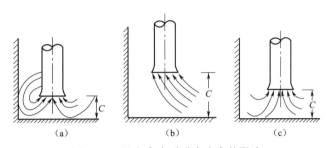

图 8.13 悬空高度对进水流态的影响

Fig. 8.13 Effect of floor clearance on flow pattern

(a) 过小；(b) 过大；(c) 适宜

(a) Too small；(b) Too large；(c) Suitable

$$L = KQ/Bh \ (\text{m}) \tag{8.6}$$

式中：h 为池中水深；K 为秒换水系数，其意义为进水池最小容积与水泵设计流量 Q 之比，视泵站供排水的重要性而定。

当 $Q < 0.5 \text{m}^3/\text{s}$，$K = 25 \sim 30 \text{s}$；当 $Q > 0.5 \text{m}^3/\text{s}$，$K = 30 \sim 50 \text{s}$。

对于轴流泵站，K 取大值，离心泵站 K 取小值。

同时应该注意，在任何情况下进水池长度应保证喇叭口中心至进水池入口距离大于 4 倍喇叭管进口直径，即 $L > 4D_L$。

5. 喇叭口淹没深度 h_s

进水池内的水位是影响水面涡形成的最主要因素。将刚出现Ⅱ型涡时的喇叭口淹没深度定义为临界淹没深度。为保证进水池不发生水面吸气涡，喇叭口淹没深度必须大于临界淹没深度。当然，淹没深度也不能过大，以免导致水泵安装高程过低，增加进水池的挖深、增加土建投资。

实际上，进水池各部分尺寸之间是相互影响、相互制约的。临界淹没深度的确定与多种因素有关，进水池的宽度及后壁距也在不同程度上影响着临界淹没深度。图 8.14 所示为一试验所得后壁距与临界淹没深度的关系曲线。由图可以看到，后壁距从 $0.5D_L$ 增至 $1.25D_L$ 时，临界淹没深度从不到 0.5m 增至 2.0m，影响是很显著的。一般地说，下列因素要求较大的临界淹没深度：

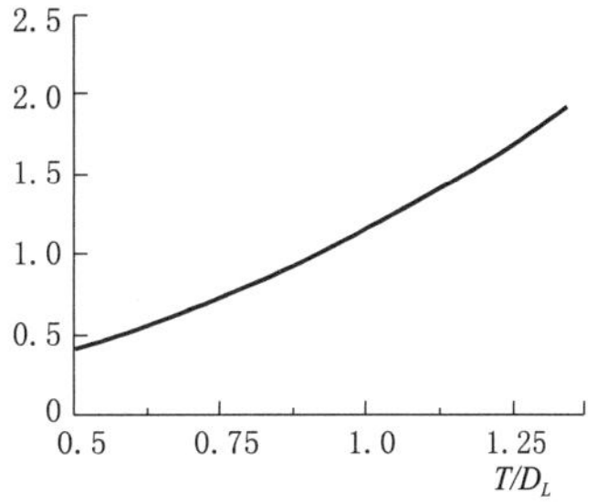

(1) 进水池中流速较大、吸水管管口流速较大。

(2) 悬空高度较小，池宽较小，后壁距较大。

图 8.14 后壁距对临界淹没深度的影响

Fig. 8.14 The influence of backwall clearance on critical submergence

临界淹没深度还与喇叭管的安装方式有关，根据试验结果，水平安装的喇叭管所需的淹没深度最大。图 8.15 给出了进水喇叭管垂直、倾斜和水平安装方式下的淹没深度 h_s。

《泵站设计规范》（GB 50265—2010）推荐：

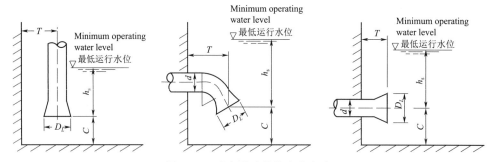

图 8.15　进水喇叭管的安装方式

Fig. 8.15　Installation method of inlet bellmouth

(a) 垂直安装；(b) 倾斜安装；(c) 水平安装

(a) Vertical installation；(b) Inclined installation；(c) Horizontal installation

（1）喇叭管垂直布置时，$h_s > (1.0 \sim 1.25) D_L$。

（2）喇叭管倾斜布置时，$h_s > (1.5 \sim 1.8) D_L$。

（3）喇叭管水平布置时，$h_s > (1.8 \sim 2.0) D_L$。

对立式轴流灌溉泵站，因其常在低水位下运行，h_s 可取大值；对立式轴流排涝泵站，因其常在进水位较高的情况下运行，h_s 可取小值。为保证不发生汽蚀，任何情况下，h_s 不得小于 0.5m。

6. 进水池后壁平面形状对流态的影响

矩形进水池 [图 8.16 (a)] 结构简单、施工方便，在中小型泵站得到广泛应用。矩形

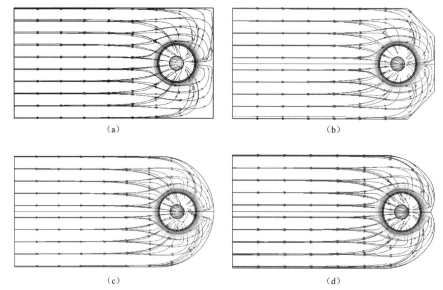

图 8.16　进水池后壁平面形状对流态的影响

Fig. 8.16　Influence of sidewall shape of sump on flow pattern

(a) 矩形；(b) 多边形；(c) 半圆形；(d) 平面对称蜗壳形

(a) Rectangle shape；(b) Polygon shape；(c) Semicircular shape；

(d) Planar symmetrical volute shape

进水池的两角处易形成漩涡，增加了进水池的水力损失。多边形进水池［图8.16（b）］在矩形进水池的基础上去掉两只角，消除两角的漩涡，平面形状更接近水流流线。半圆形进水池［图8.16（c）］边壁形状与水流流线也较接近，但如果安装位置不当或淹没深度较小时，易产生漩涡。平面对称蜗壳形（或称为W形）［图8.16（d）］最符合水流流线要求，水力损失最小且不易产生漩涡，这种进水池目前应用较广。试验结果表明，平面对称蜗壳形进水池的水力损失系数最小，水泵装置效率最高，应优先考虑采用。

8.4 进 水 流 道

开敞式进水池一般适用于中小型泵站。大型立式轴流泵和导叶式混流泵的进水结构如仍采用喇叭口直接从进水池吸水，则按照上一节要求，进水池的尺寸往往很大，将使得水泵支承结构尺寸随之加大。同时，大型泵站流量较大，对流态的要求更高，但进水池来流方向与水泵进口水流方向要成90°的转弯，因此很难保证水泵进口有均匀的流场条件。所以，为了节省工程量和改善水力条件，大型立式（斜式）泵站将进水池和吸水管合二为一，采用专门设计的进水流道从前池中取水。

8.4.1 进水流道作用及水力设计要求

1. 进水流道作用

进水流道有多种形式，各种不同的进水流道尽管形式不一，但都是泵站前池与水泵叶轮室之间的过渡段，其作用都是使水流在从前池进入水泵叶轮室的过程中更好地转向和加速，以尽量满足水泵叶轮对叶轮室进口所要求的水力设计条件。

2. 进水流道设计要求

水泵特性都是在专用的水泵试验台上经过性能试验获得的。根据相关的国家试验标准，在受试泵叶轮室之前必须配备不少于15倍管道直径的平直管段。这是为了保证受试泵的进口流态最大限度地满足水泵叶轮的水力设计条件。泵站进水流道与实验室标准管道所提供的进水流场不可避免地存在差别，进水条件的变化必然引起泵装置中水泵工作状态的变化。进水流态不良不仅会降低水泵效率，也会降低水泵的汽蚀性能。因此，进水流道的水力设计，将直接影响水泵的工作状态，进水流态越差，对水泵实际性能的影响就越大。可见，进水流道是水泵装置的一个重要组成部分。

进水流道的设计应满足以下要求：

（1）流道型线平顺，各断面面积沿程变化应均匀合理。

（2）出口断面处的流速和压力应比较均匀。

（3）进口断面处流速宜取0.8～1.0m/s。

（4）在各种工况下，流道内不应产生涡带。

（5）进口宜设检修设施。

（6）施工方便。

为水泵进口提供良好的流态、保证水泵机组稳定高效运行是进水流道的首要任务。在上述要求中，第（4）项要求应确保满足，第（2）项要求应尽可能满足；第（1）项要求实际上是能够得到良好进水流态的必要条件。在此基础上，可适当兼顾减少土建投资及施工方便等其他方面的要求。

在相关水力优化设计研究中，还采用泵进口流速分布均匀度和入泵速度加权平均角度两个参数评判进水流道优劣。

8.4.2 常用进水流道简介

常用的进水流道有肘形进水流道、斜式进水流道、钟形进水流道、簸箕形进水流道等单向进水流道，此外还有与贯流泵装置、双向抽水泵装置等配套使用的特殊形式的进水流道。

1. 肘形进水流道

肘形进水流道（图 8.17）适用于立式轴流泵或导叶式混流泵，其形状与水轮机尾水管相近，在我国的大型泵站中应用最早、最为广泛。肘形进水流道的特点是高度较大而宽度较小，可得到很好的水力性能；其缺点是挖深较大。传统的肘形进水流道水力设计采用建立在一维流动理论基础上的平均流速法。这种方法的主要缺点是只考虑流道内平均流速的变化，而未考虑流道内流速的分布，因而不能按照要求的流场要求设计流道。一维设计理论中存在的问题促使人们对进水流道的水力设计方法展开了深入的研究，从而推动了进水流道三维优化水力设计理论的发展。

2. 斜式进水流道

斜式进水流道（图 8.18）与斜式轴伸泵装置配套应用，按水泵轴线与水平线的夹角一般分为 45°、30°和 15°三种型式，以适应不同的水泵装置扬程。斜式进水流道的水力性能优异、形状简单、土建投资省，但需解决好齿轮箱和轴承制造质量等问题。20 世纪 80年代以来斜式轴伸泵在我国逐步得到应用。1986 年，上海水泵厂从日本荏原公司引进 45°斜式轴伸泵装置全套技术，为内蒙古红圪卜泵站制造了 6 台直径为 2.5m 的斜式轴流泵。我国自行研制开发的 15°和 30°斜式轴伸泵装置也分别运用于湖南铁山嘴排涝站和江苏新夏港泵站。此外，太湖流域综合治理工程中的浙江盐官泵站和上海太浦河泵站也采用了斜式轴伸泵装置。

图 8.17 肘形进水流道

Fig. 8.17 Elbow-shaped inlet passage

图 8.18 斜式进水流道

Fig. 8.18 Inclined inlet passage

3. 钟形进水流道

钟形进水流道（图 8.19）的显著特点是高度较小、宽度较大，这对于站址地质条件较差的泵站具有特别重要的意义；其缺点是形状复杂，施工不便，且对流道宽度的要求非常严格，设计不当，易在流道内产生涡带。钟形进水流道早期在日本的一些大型排灌泵站应用较多，20 世纪 70 年代起，在我国的大型泵站建设中也得到了一些应用，如湖南坡头泵站、湖北新沟泵站、罗家路泵站及江苏临洪西站、皂河泵站等。

4. 簸箕形进水流道

簸箕形进水流道（图 8.20）在荷兰等欧洲国家应用广泛，大、中、小型泵站都用，已有 70 多年的历史。这种流道形状较为简单，施工方便。近几年来，簸箕形流道在我国开始得到应用。上海郊区首次将这种流道应用于小型泵站的节能技术改造，江苏的刘老涧泵站首次将这种流道应用于大型泵站，预计今后可能会得到更多应用。簸箕形进水流道在基本尺寸方面介于肘形流道和钟形流道之间，对流道宽度的要求没有钟形流道那样严格，不易产生涡带。

图 8.19　钟形进水流道

Fig. 8.19　Bell-shaped inlet passage

图 8.20　簸箕形进水流道

Fig. 8.20　Dustpan-shaped inlet passage

8.4.3　进水流道的水力设计

1. 一维水力设计方法

传统的肘形进水流道水力设计方法是典型的一维水力设计方法，其要点为：假定断面平均流速等于设计流量除以断面面积；以沿流道断面中心线的各断面平均流速光滑变化为目标。

2. 三维优化水力设计方法

随着计算流体动力学的迅速发展，以三维湍流流动理论为基础的进水流道优化水力设计方法于 20 世纪 90 年代初被提出并逐步得到了实际应用。通过给定一系列不同的进水流道边界、采用三维湍流流动数值模拟的方法完成一系列相应的流场计算，根据流场计算结果逐一地修改、优化进水流道各几何参数，成为泵站进水流道优化水力设计的基本思路。

8.4.4　进水流道主要尺寸推荐值

大量的数值计算和模型试验数据表明，进水流道的水力设计准则与流道内的基本流态密切相关，每个几何参数的确定都服从于它们对进水流态的影响程度。

根据试验与 CFD 优化水力计算成果，推荐的单面进水流道主要控制尺寸列于表 8.1。

叶轮中心高度是流道设计中最重要的参数，此高度越大，水泵进口的流态越好，但所需的泵站土建投资也越多。这一矛盾对肘形进水流道尤为突出。表 8.1 中的推荐值兼顾了进水流态和土建投资两方面的要求。

表 8.1   进水流道主要尺寸推荐值

Table 8.1   **Recommended dimensions of inlet passages**

类型 type	叶轮名义高度 nominal height of impeller H_w/D_0	流道宽度 passage width B/D_0	流道长度 passage length L/D_0
15°斜式流道 15° inclined passage	0.7～0.9	2.3～2.5	3.0～4.0
30°斜式流道 30° inclined passage	0.8～1.0	2.3～2.5	3.0～4.0
45°斜式流道 45° inclined passage	1.0～1.2	2.3～2.5	3.2～4.0
肘形流道 elbow-shaped passage	1.6～1.8	2.3～2.5	3.5～4.0
钟形流道 bell-shaped passage	1.3～1.4	2.8～3.0	3.5～4.0
簸箕形流道 dustpan-shaped passage	1.5～1.6	2.5	3.5～4.0

    流道直线段的长度和宽度对水泵叶轮室进口的流场影响较小，长度一般可视泵房上部顺水流方向结构布置的要求确定；流道宽度则可根据机组中心距等布置方面的要求确定。

Chapter 8   Inlet Structures

Inlet structures of pumping station mainly include inlet headwork, inlet culvert gate, open or dark diversion channel, forebay or collecting well, suction sump or suction chamber, etc. Large and medium pumping stations also include inlet passages. The layout of inlet structures and hydraulic design are reasonable or not, which directly affects the working performance of pumps and the efficiency of pumping stations. In order to ensure safe and economic operation of the pumping station, the inlet structures of the pumping station should meet the following requirements, such as sufficient inlet capacity, good inlet flow pattern, stability and saving civil investment.

8.1   Inlet and diversion structures

8.1.1   Inlet structures

The structures that use culverts to draw water from the water source are called inlet heads. There are many structural forms, including gravity type, caisson type, frame type, cantilever type, bottom trough type and tunnel type.

There are inlet culverts, sluices, open inlet and other structures which take water from the bank of water source. For inlet in sediment-laden rivers, favorable position should be selected, surface water with little sediment content should be taken, bottom water with high sediment content should be diverted by diversion facilities, and sediment sink should be set at proper position of diversion channel.

8.1.2   Diversion structures

When it is really difficult to build a pumping station near the water source or the bank of the drainage area, a diversion structure should be set up, which can take the form of open diversion channel or use existing rivers and pressure culverts. The main functions of the pumping station diversion channel as the open channel connecting the water source (or drainage area) and the pump house are presented as follows:

(1) The stream current smoothly into the forebay.

(2) It will be avoid direct near between pump house and water source, simplify pump house structure and facilitate construction.

(3) It will be provide conditions for sediment settling in sand-bearing rivers.

Requirements for diversion channels are shown below:

(1) Be able to transfer the discharge required by the pumping station at any time.

(2) It can be adapt to the variation of pumping discharge in pumping station.

(3) Less hydraulic loss.

1. Route selection of diversion channel

Route selection of diversion channel should be determined by technical and economic comparison according to selected inlet and pump house location, combined with various factors such as topographic and geological conditions, construction conditions and balance of excavation and filling. Channel lines should avoid sections with complex geological structure, strong permeability and possibility of collapse. Channel body should be located on excavated foundation and occupy less cultivated land. In order to reduce the amount of work, the channel lines should be straight. When setting bends, the radius of soil channel bends should not be less than 5 times of the width of water surface of the channel, the radius of stone channel and lining channel bends should not be less than 3 times of the width of water surface of the channel. There should be a straight line between the end point of the bends and the inlet of the forebay, and the length should not be less than 8 times of the width of the water surface of the channel.

2. Types of diversion channels

(1) Automatic regulation, the elevation of the tail top of the diversion channel is basically the same, and the top line of the channel is higher than the maximum water level line of the water source. Therefore, no matter how much discharge passes through the channel, the water level will not exceed the top of the embankment and overflow will occur, so there is no need to set control structures for the diversion channel. Its advantage is that the channel has a certain volume, which can slow down the large fluctuation of water level in the channel when the pump starts and stops. Its disadvantage is large excavation volume and the pump house should have flood control measures.

(2) Without automatic regulation, the top line of the levee of the diversion channel is parallel to the bottom slope line of the channel. When the design flow is passed through the channel, the stream current in the channel is uniform, and the water surface line is parallel to the bottom of the channel, and the water depth is equal to the normal water depth. When the pumping discharge is less than the design discharge, the stream current in the channel is in non-uniform flow state and the water surface curve is a backwater profile. To ensure that the diversion channel does not overflow, overflow structures are usually set at the end of the diversion channel or control gates are set at the head of the channel. Its advantages are: less excavation, suitable for long-distance diversion channel; no flood control facilities need to be set before pump house. Its disadvantage is that the structure with controlled water level needs to be increased.

3. Design of diversion channel section

The design method of diversion channel is basically the same as that of general water conveyance channel, i. e. it is designed according to uniform flow and checked according to conditions of no scouring and no silting.

4. Impact of pump start-up and stop on flow pattern in diversion channel

When the pump is started, the water surface in the channel will drop, while when the pump is stopped, the water surface in the channel will backup. Water surface drops and backup heights occur in the form of waves. Reverse falls occur when dropping and rise occurs when backup is high (Fig. 8. 1) . When determining the submergence depth required for the inlet of the suction pipe of the pump, the influence of the water surface drop (i. e. negative surge) caused by the start-up of the last unit should be considered; when determining the top elevation of the channel, the influence of the water surface backup (i. e. positive surge) caused by the sudden stop of the unit should be considered.

The depth of the negative surge can be approximated by the following formula:

$$\Delta h_n = 2 \frac{Q_1 - Q_0}{B \sqrt{g h_0}} \ (\text{m}) \tag{8.1}$$

Where, Q_1 and Q_0 are the flow rate in the channel before and after the start-up of the pump unit, m^3/s; B is the average surface width in the channel, m; h_0 is the depth of water in the channel before starting or stopping the pump unit, m.

The depth of the positive surge can be approximated by the following formula:

$$\Delta h_p = \frac{(v_0 - v') \sqrt{h_0}}{2.76} - 0.01 h_0 \ (\text{m}) \tag{8.2}$$

Where, v_0 is the velocity at the end of the diversion channel before sudden stop, m/s; v' is the velocity at the end of the diversion channel after sudden shutdown, m/s.

The positive surge caused by backup of water surface may affect the flood control elevation of the levee top of the diversion channel and the pump house. The negative surge caused by water surface drop may reduce the submergence depth of the pump inlet, cause cavitation of the pump and unit vibration, and even affect the normal operation of the pump.

In conclusion, the operation characteristics of pumping station should be fully reflected in the design of diversion channel, i. e. the requirement that diversion channel can create good flow conditions for safe operation of pump. To determine the design flow of diversion channel, the diversion channel should be kept in constant velocity flow or backwater slightly, and under no circumstances should the water surface drop be excessive. The length and longitudinal slope of the channel should also be comprehensively selected through comprehensive calculation of project cost and operation cost.

8. 2   Forebay

Forebay is a structure connecting the diversion channel with the suction sump. Its function is to smoothly and evenly convey the stream current from the diversion channel to the suction sump, providing good water-suction conditions for the pump, and having a certain volume to reduce the change of water level when the pump starts and stops. Fig. 8. 2 shows the connection of the forebay of a typical multi-unit pumping station to the diversion channel and inlet sump.

8. 2. 1 Function and type of forebay

According to the position of forebay and suction sump, it can be divided into front forebay and side forebay.

The center line of the front inlet forebay coincides with that of the diversion channel and suction sump, and the flow direction is basically the same [Fig. 8. 3 (a)]. The flow direction of the side inlet forebay is orthogonal or oblique to that of the suction sump [Fig. 8. 3 (b)].

In the front and side forebays, it can be divided into two types, with and without piers. The function of the pier is:

(1) Divert water diversion to make the stream current stable and uniformly enter each water suction sump, especially to avoid large area shaking, backflow and swirl of the stream current in the forebay during asymmetric start-up.

(2) If the forebay divergence angle is too large, it can reduce the actual divergence angle of stream current and shorten the sump length.

(3) Piers can also be used as sluice piers and working piers.

8. 2. 2 Unfavorable flow pattern and hydraulic design requirements of forebay

1. Poor flow pattern in forebay

(1) Front forebay. The front forebay is designed reasonably. When all or part of the pump is started symmetrically, the flow pattern is better. The front forebay should be adopted as far as possible in all pumping stations. However, when some pumps are operated asymmetrically, or when the water level of the forebay is low, the flow rate of the pumping station is large, the diffusive angle of the forebay is too large, and the flow pattern of the inlet diversion channel is poor, the following undesirable flow patterns will also occur in the front inlet forebay:

1) Mainstream deflection, backflow and dead water zone occur on one side forebay, and some suction sumps become side inlet sumps.

2) Mainstream is centered. Dewall backflow occurs on both sides of the forebay. The middle suction sump is overwashed by the mainstream and the two sides sump become side suction sumps as shown in Fig. 8. 5 (a).

3) An enlarged folding flow occurs at the inlet of the forebay, and the flow pattern in the forebay is very disordered.

The first and second flow patterns are easy to occur when the forebay divergence angle is large, the water level is low and the flow rate of the pumping station is large. The third flow pattern is caused by diversion channel inflow with folding state.

(2) Side forebay. The stream current in the side inlet forebay is usually in a bad state. Because the flow direction of water in the forebay needs to change, it is easy to produce side wall off-flow and large area back-flow, and uneven approaching velocity distribution is generated before the inlet sump of pumping station, which leads to swirls on the back surface of the head of the partition pier of the inlet sump of each pump station unit, the flow direction in the sump is biased to one side, and water surface vortices are generated near the suction port of the pump, etc. As shown in Fig. 8.5 (b).

2. Hydraulic design requirements of forebay

The experimental study shows that the undesirable flow pattern in the forebay will seriously affect the flow pattern in the sump, which will lead to the decrease of energy and cavitation performance of the pump. At the same time, large-area backflow will also cause local deposition in the forebay, which will further aggravate the development of the undesirable flow pattern. The hydraulic design of the forebay requires that the flow of water be smooth and the divergence be smooth without wall separation, backflow or whirlpool. Meanwhile, the civil investment should be considered as much as possible.

8.2.3   Dimension of forebay

1. Divergence angle α

The divergence angle α not only affects the flow state of the forebay, but also affects the engineering quantity. If the α is too large, the length of forebay will be short, and the engineering amount will be small, but the stream current will not diffuse enough, which will easily lead to detachment and backflow; if the α is too small, the stream current in the forebay will diffuse smoothly, but the sump length will grow.

The stream current has a natural divergence angle when it diffuses in the gradient section, i. e. the critical divergence angle without detached backflow. Its value can be calculated according to the following semi-empirical and semi-theoretical formula:

$$\tan \frac{\alpha}{2} = 0.065 \frac{1}{\sqrt{Fr}} + 0.107 \tag{8.3}$$

$$Fr = \frac{v}{\sqrt{gh}}$$

When $Fr = 1$, i. e. the stream current is in the critical state between torrent flow and slow flow, $\alpha = 20°$. Since the flow in the forebay is usually slow, Fr number is smaller than 1, the divergence angle α can be greater than 20°. According to practical engineering experience, the divergence angle of the forebay is generally range from 20° to 40°.

2. Length L

The length of the forebay can be calculated from the bottom width of the diversion channel end, the total width of the suction sump and the selected divergence angle of the forebay:

$$L = \frac{B-b}{2\tan\frac{\alpha}{2}} \quad (\text{m}) \tag{8.4}$$

3. Bottom slope of forebay i

Because of the requirement of submergence depth of pumps, the elevation of the bottom of suction sump is generally lower than that of the bottom of diversion channel. Therefore, the bottom of forebay is a slope, which connects in the elevation direction and has a slope of

$$i = \frac{\Delta H}{L} \tag{8.5}$$

Where, ΔH is the difference between the bottom elevation of the channel at the end of the diversion channel and the bottom elevation of the suction sump.

According to the relevant test data, the forebay bottom slope has certain influence on the inflow flow pattern. Fig. 8. 6 shows the relationship between the inlet drag coefficient of the inlet pipe and the forebay bottom slope. It can be seen from the diagram that the bigger the slope of the forebay bottom, the bigger the drag coefficient. When i is smaller than 0. 3, the variation of the drag coefficient is small; when i is greater than 0. 3, the variation of the drag coefficient is large. Therefore, i should be selected in the range from 0. 2 to 0. 3. When i is smaller than 0. 3, the front section of the forebay can be made horizontal and the rear section near the suction sump can be made slope in order to save the quantity of work.

8. 2. 4 Improvement of flow pattern in forebay

The flow pattern in the forebay with front inlet is good, so rectification measures are usually not necessary. Only when the diffusion angle of the forebay is too large or the startup is often asymmetrical, diversion piers, sills and columns should be set to improve the flow pattern in the forebay.

The flow pattern in the forebay with side inlet is poor, so proper rectification measures should be taken generally. Fig. 8. 7 shows how the inlet pattern is improved by column and sill rectification in a pumping station side inlet forebay. Tests and on-site operation of pumping station show that the large area of backflow area of the original scheme is destroyed after rectification with column and sill in proper position, the approaching velocity distribution in front of the suction sump of two pumping stations is improved, and the front uniform inflow of most of the suction sumps is realized. From the overall effect, the operation conditions of the pumps have been greatly improved.

8. 3 Suction sump

8. 3. 1 Function and design requirements of sump

Open sump is the structure of water supply pump or inlet pipe which absorbs water directly. Its main functions are:

(1) Further adjust the inflow from the forebay to provide good inflow conditions for the pump inlet.

(2) Cut off stream current while overhauling pump or inlet pipe.

(3) Intercept the dirt in the inflow.

The pump impeller is designed on the assumption that the inlet flow is uniform. For vertical axial flow pumps and guide vane mixed flow pumps, the sump is the lower floor of the wet-pit pump house, also known as the pump chamber. Because the impeller casing of the pump is close to the suction horn, the guide function of the suction horn is poor, and the flow pattern in the sump has a great influence on the performance of the pump. For horizontal pumps with water pipes, since the inlet pipes can automatically adjust the uneven inlet flow pattern, the flow pattern in the suction sump is slightly lower than that of the vertical pump. However, the harmful vortices in the suction sump will still affect the stable and safe operation of the pump.

Therefore, the design requirements for suction sump are as follows:

(1) The flow pattern in the sump is good without obvious swirl back flow.

(2) Meet the requirement of inlet of pumping station.

(3) It is convenient for dredging, management and maintenance.

8. 3. 2   Geometric parameters of suction sump

The main geometrical parameters of rectangular suction sump include width B of suction sump, floor clearance C of bellmouth, distance T of backwall clearance, length L of sump and submergence depth h_s of bellmouth, as shown in Fig. 8. 8. In the Fig. , D_L is the diameter of the inlet bellmouth. At present, most of the geometric parameters of the suction sump are represented by D_L as the basic parameters.

8. 3. 3   Basic flow pattern in suction sump

The basic characteristic of the flow in suction sump is that the water flows from all sides into the bellmouth. As shown in Figure 8. 9, when the pump is running, part of the water flows into the bellmouth from the front, part of the water flows into the bellmouth from both sides, and part of the water flows around both sides and into the bellmouth from the back.

8. 3. 4   Bad flow pattern and vortex in the suction sump

The bad flow patterns in the suction sump are mainly uneven flow velocity distribution on various sections and various forms of vortices. Among them, the most harmful to the operation of the pump is the vortex entering the pump.

1. Type of vortex in the sump

In the case of improper design of the suction sump or the submerged depth of the suction pipe of the pump, the swirl may occur in the suction sump.

According to the location of the vortex, the vortex can be divided into three types:

bottom-attached vortex, wall-attached vortex (Coanda vortex) and water surface vortex, as shown in Fig. 8. 10.

According to the absorption of water surface vortices, they are usually divided into four types, as shown in Fig. 8. 11.

Type Ⅰ vortex: the stream current rotates slowly, only a shallow funnel is formed on the water surface, and air has not been brought into the pump.

Type Ⅱ vortex: the stream current rotates faster, the water funnel is deep, and the air has been brought into the pump intermittently, which has only a slight effect on the pump performance.

Type Ⅲ vortex: the stream current rotates rapidly, the water funnel has been put into the suction pipe, and the air enters the pump continuously, which has a serious impact on the operation of the pump.

Type Ⅳ vortex: The stream current rotates sharply around the suction pipe. The center of the vortex is consistent with the center of the suction pipe. A large amount of air enters the pump. The pump unit produces severe vibration so that it cannot operate.

When the depth of submergence is large, only vortices of type Ⅰ and Ⅱ may occur. When h_s decreases, funnel-like vortex of type Ⅲ occurs, and when h_s further decreases, vortices of type Ⅳ occurs. Vortices of type Ⅰ–Ⅲ are called local vortices and vortices of type Ⅳ are called cylindrical or concentric vortices.

2. Vortex prevention and control measures

The vortices in the suction sump affect the operation of the pump to some extent. Therefore, besides rectification measures such as column or sill in the forebay, the vortices can be prevented by setting baffles and diversion cones in the suction sump (Fig. 8. 12).

8. 3. 5   Determination of Sump Size

At present, the determination of the geometric size of the suction sump is mostly dependent on the test results, and the test data obtained are often inconsistent due to the differences in the test conditions. With the rapid development of computational fluid dynamics (CFD), theoretical calculation method has been used to study and solve the hydraulic design problem of the suction sump in China.

1. Suction sump width B

The width of the suction sump B is too small, which will accelerate the flow rate and increase the head loss in the sump, increase the streamline curvature when the flow converges to the horizontal direction of the bellmouth, and easily induce eddies. If the width of the suction sump B is too large, it will not only increase the investment in civil engineering, but also reduce the guiding effect of the sump. It is easy to form deviations and backflows in the sump and generate swirls. According to the analysis of the basic flow pattern of the suction sump, the stream current enters the bellmouth from all around. If the sump width is too small, it will inevitably affect a part of the stream current to enter the pump smoothly from both sides and back of the bellmouth. Therefore, a certain sump width is needed, but the sump width need not be too large, otherwise it will increase the civil investment.

In determining the sump width, besides the hydraulic conditions, the requirements of unit installation and maintenance should also be considered. Generally, $B = (2-3) D_L$ is required. In the case of one sump and multiple pumps, in order to reduce the interaction between the pumps, the distance between two adjacent pumps can be appropriately increased, and $(3-4) D_L$ is taken.

2. Floor clearance C

The floor clearance refers to the distance from the inlet of the bellmouth to the sump floor, and its value has a significant impact on the flow pattern and civil investment near the bellmouth. The stream current in the suction sump enters the bellmouth from around, and a suitable floor clearance is essential to form such a flow and to make the stream current enter the bellmouth essentially uniformly. Excessively high floor clearance not only increases the depth of excavation and investment, but also may form a one-sided water inlet of bellmouth, resulting in uneven distribution of flow rate and pressure at the inlet of the pump, reducing the energy performance and cavitation performance of the pump, and even sometimes producing a bottom-attached eddy or wall-attached eddy. If the floor clearance is too small, the cylindrical surface under the bellmouth will be compressed, resulting in the streamlines inflow into the bellmouth being too curved, the hydraulic loss at the inlet of the bellmouth will increase sharply, and the velocity and pressure distribution at the inlet of the pump will also be uneven, which will increase the possibility of swirls. The effect of floor clearance on inflow flow is shown in Fig. 8. 13.

Based on the relevant research results at home and abroad, it is suggested that the floor clearance $C = (0.6-0.8) D_L$.

3. Back wall distance T

Back wall distance refers to the distance from the center of the suction tube to the back wall of the suction sump. By analyzing the basic flow pattern of the suction sump, it can be noticed that part of the stream current enters the bellmouth from the back of the bellmouth, so there must be some space on the back wall. Small rear wall spacing will result in uneven flow and large water loss of the bellmouth. Excessive back wall spacing is not only unnecessary, but also increases the degree of freedom of stream current in the back wall space, thus increasing the possibility of suction swirls. The results show that the larger the backwall distance, the greater the submergence depth required.

Reasonable back wall distance should also meet the requirements of inlet tube installation. It is suggested that $T = (0.8-1.0) D_L$.

4. Suction sump length L

The length of the sump refers to the distance from the inlet of the sump to the center of the suction pipe. The length of the suction sump should be able to adjust the flow rate in the sump to make it uniform and maintain a certain volume to avoid sharp drop of water level. Determine according to

$$L=KQ/Bh \ \ (m) \tag{8.6}$$

Where, h is water depth in sump; K is factor, which means the ratio of the minimum volume of the suction sump to the design flowrate Q of the pump, depends on the importance of water supply and drainage of the pumping station.

If $Q<0.5m^3/s$, $K=25-30s$; if $Q>0.5m^3/s$, $K=30-50s$.

For axial flow pumping station K is taken as large value, while centrifugal pumping station K is taken as small value.

At the same time, it should be noted that in any case, the length of the suction sump should ensure that the distance between the center of the bellmouth and the inlet of the suction sump is greater than 4 times that of the inlet of the bellmouth tube, i. e. , $L>4D_L$.

5. Bellmouth submergence depth h_s

The water level in the suction sump is the most important factor affecting the formation of free water surface vortices. The submerged depth of the bellmouth when the vortex of type II just appears is defined as the critical submerged depth. In order to ensure that the suction vortices on the water surface do not occur in the suction sump, the submerged depth of the bellmouth must be greater than the critical submerged depth. Of course, the submergence depth should not be too large, so as not to cause the pump installation elevation too low, increase the depth of the suction sump, increase civil investment.

In fact, the sizes of each part of the suction sump affect and restrict each other. The determination of critical submergence depth is related to a variety of factors, and the width of the suction sump and the distance between the back walls also affect the critical submergence depth to varying degrees. Fig. 8. 14 shows a curve of the relationship between the backwall distance and the critical flooding depth obtained from the experiment. As can be seen from the Fig. , the critical submergence depth increases from less than 0. 5m to 2. 0m when the distance between the back wall increases from 0. 5 D_L to 1. 25 D_L , and the effect is very significant. Generally speaking, the following factors require greater critical flooding depth:

(1) The flow rate in the suction sump is large, and the flow rate at the inlet of the suction pipe is large.

(2) The floor clearance is smaller, the sump width is smaller, and backwall distance is larger.

The critical submergence depth is also related to the installation mode of the bellmouth. According to the test results, the horizontal installation of the bellmouth requires the greatest submergence depth. Fig. 8.15 shows the submergence depth h_s under vertical, inclined and horizontal installation of the inlet bellmouth.

Design code for Pumping station (GB 50265—2010) recommends:

(1) When the bellmouth is arranged vertically, $h_s > (1.0 - 1.25) D_L$.

(2) When the bellmouth is arranged obliquely, $h_s > (1.5 - 1.8) D_L$.

(3) When the bellmouth is arranged horizontally, $h_s > (1.8 - 2.0) D_L$.

For vertical axial flow irrigation pumping station, h_s can be taken as a large value because it usually operates at low water level; for vertical axial flow drainage pumping station, h_s can be taken as a small value because it often operates at high water level. To ensure that no cavitation occurs, the h_s shall not be less than 0.5m under any circumstances.

6. Influence of sidewall shape of sump on flow pattern

Rectangular sump [Fig. 8.16 (a)] is widely used in small and medium pumping stations for its simple structure and convenient construction. Swirls are easy to form at the corners of rectangular suction sump, which increases the hydraulic loss of suction sump. Polygonal sump [Fig. 8.16 (b)] removes two corners on the basis of rectangular sump and eliminates the whirlpools at the two corners. The sidewall shape is closer to the streamline. The shape of the side wall of the semicircular sump [Fig. 8.16 (c)] is also close to the streamline, but if the installation position is improper or the submergence depth is small, swirls are likely to occur. The planar symmetrical volute shell shape (or W-shape) [Fig. 8.16 (d)] is most suitable for streamline, has the least hydraulic loss and is not easy to generate vortices. This type of inlet tank is widely used at present. The test results show that the hydraulic loss coefficient of the planar symmetrical volute-shell inlet tank is the smallest and the efficiency of the pump unit is the highest.

8.4 Inlet passage

Open sump is generally suitable for small and medium-sized pumping stations. If the inlet structure of large vertical axial flow pump and guide vane mixed flow pump still uses flare to absorb water directly from the suction sump, the size of the suction sump is usually large according to the requirements of the previous section, which will increase the size of the pump support structure. At the same time, large pumping stations have a large flow rate and higher convection requirements, but the inflow direction of the suction sump and the flow direction of the pump inlet are 90 degrees turning, so it is difficult to ensure uniform flow conditions at the pump inlet. Therefore, in order to save engineering quantity and improve hydraulic conditions, large vertical (inclined) pumping stations combine suction sump and suction pipe and use specially designed inlet passages to take water from the forebay.

8. 4. 1　Function of inlet passage and hydraulic design requirements

1. Function of inlet passage

There are many types of inlet passages. Although different types of inlet passages are different, they are transitional sections between the pumping station forebay and the pump impeller chamber. Their functions are to make the stream current better turn and accelerate during the process of entering the pump impeller chamber from the forebay, so as to meet the hydraulic design conditions required by the pump impeller for the impeller chamber inlet as far as possible.

2. Hydraulic design requirements of inlet passage

Pump characteristics are obtained by performance tests on a dedicated pump test stand. According to the relevant national test standards, the impeller chamber of the pump under test must be equipped with a straight section with not less than 15 times the diameter of the pipe. This requirement is made to ensure that the inlet flow pattern of the pump under test meets the hydraulic design conditions of the pump impeller to the maximum extent. Inlet passage of pumping station is inevitably different from that provided by standard piping in laboratory. The change of inlet condition will inevitably cause the change of pump operating state in pump unit. Poor inlet flow not only reduces pump efficiency, but also reduces pump cavitation performance. Therefore, the hydraulic design of inlet passage will directly affect the working state of the pump. The worse the inlet flow, the greater the impact on the actual performance of the pump. It can be seen that the inlet passage is an important part of the pump unit.

The design of inlet passage shall meet the following requirements:

(1) The flow passage is smooth and the area of each section should change uniformly and reasonably along the passage.

(2) The flow rate and pressure at the outlet cross section should be uniform.

(3) The flow rate at the inlet section should be 0. 8 – 1. 0m/s.

(4) Vortex shall not occur in the flow passage under different operating conditions.

(5) Inspection gate should be set at the entrance.

(6) Construction is convenient.

Providing a good flow pattern for the inlet of the pump and ensuring the stable and efficient operation of the pump unit are the primary tasks of the inlet passage. Among the above requirements, the requirements of item (4) shall be met, the requirements of item (2) shall be met as far as possible, and item (1) shall in fact be a necessary condition for a good inlet flow pattern. On this basis, other requirements, such as reducing civil investment and facilitating construction, can be given due consideration.

In the relevant hydraulic optimum design research, 2 parameters, uniformity of velocity distribution at pump inlet and weighted average angle of inlet pump velocity, are also used to evaluate the quality of inlet inlet passage.

8. 4. 2 Introduction to common inlet passages

Commonly used inlet passage types are elbow-shaped inlet passage, inclined inlet passage, bell-shaped inlet passage, dustpan-shaped inlet passage and other one-way inlet passage. In addition, there are special types of inlet passage matched with tubular pump device and bidirectional pump device.

1. Elbow-shaped inlet passage

The elbow-shaped inlet passage (Fig. 8. 17) is suitable for vertical axial flow pump or guide vane mixed flow pump. Its shape is similar to that of draft passage of turbine. It is the earliest and most widely used in large pumping stations in China. The elbow-shaped inlet passage is characterized by large height and small width, which results in good hydraulic performance; the disadvantage is large excavation depth. The traditional hydraulic design of elbow-shaped inlet passage uses the average velocity method based on the one-dimensional flow theory. The main disadvantage of this method is that it only considers the change of average velocity in the inlet passage, but does not consider the distribution of velocity in the inlet passage, so the flow passage can not be designed according to the required flow field requirements. Problems existing in the one-dimensional design theory prompt people to carry out in-depth research on hydraulic design method of inlet passage, which promotes the development of three-dimensional optimal hydraulic design theory of inlet passage.

2. Inclined inlet passage

Inclined inlet passage (Fig. 8. 18) is used in conjunction with inclined axle extension pump unit. According to the angle between pump axis and horizontal line, it is generally divided into three types of 45°, 30° and 15°, in order to adapt to different pump unit head. Inclined inlet passage has excellent hydraulic performance, simple shape and low civil investment, but problems such as manufacturing quality of gearbox and bearing need to be solved. Inclined axle-extension pump has been gradually applied in China since 1980s. In 1986, Shanghai Pump Company introduced a complete set of technology of 45° inclined axle-extension pump unit from Japanese Ebara Company, and manufactured 6 inclined axle-flow pumps with a diameter of 2. 5m for Honggebu pumping station in Inner Mongolia. The 15° and 30° diagonal axle-extension pump units independently developed in China are also used in Tieshanzui drainage station in Hunan Province and Xinxiagang pumping station in Jiangsu Province. In addition, inclined axle-extension pump units are also used in Yanguan pumping station in Zhejiang Province and Taipu River pumping station in Shanghai in the comprehensive treatment project of Taihu Lake Basin.

3. Bell-shaped inlet passage

The bell-shaped inlet passage (Fig. 8. 19) is characterized by its small height and large width, which is of particular importance for pumping stations with poor geological conditions at the station site. Its disadvantage is complex shape, inconvenient construction, strict requirements on the width of the inlet passage, improper design and easy to generate vortices in the inlet passage. Bell-shaped inlet passage was widely used in some large irrigation and drainage pumping stations in Japan in the early days. Since 1970s, it has also been used in the construction of large pumping stations in China, such as Potou pumping station in Hunan Province, Xingou pumping station and Luojialu pumping station in Hubei Province, Linhong West pumping station and Zaohe pumping station in Jiangsu Province .

4. Dustpan-shaped inlet passage

Dustpan-shaped inlet passage (Fig. 8. 20) has been widely used in large, medium and small pumping stations in European countries such as the Netherlands for more than 70 years. The shape of the inlet passage is simple and the construction is convenient. In recent years, dustpan inlet passage has been applied in China. This inlet passage is used for the first time in the energy-saving technical renovation of small pumping stations in the suburbs of Shanghai and for the first time in Liulaojian pumping station in Jiangsu. More applications are expected in the future. The dustpan-shaped inlet passage is between elbow-shaped and bell-shaped inlet passage in basic dimensions. The width of the inlet passage is not as strict as that of bell-shaped inlet passage and vortex bands are not easy to produce.

8. 4. 3   Hydraulic design of inlet passage

1. One-dimensional hydraulic design method

The traditional hydraulic design method of elbow-shaped inlet passage is a typical one-dimensional hydraulic design method. Its main points are as follows: assume that the average flow velocity of the section is equal to the design discharge divided by the section area; take the smooth change of the average flow velocity of each section along the center line of the inlet passage section as the goal.

2. Three-dimensional optimum hydraulic design method

With the rapid development of computational fluid dynamics, the hydraulic design method for inlet passage optimization based on three-dimensional turbulent flow theory was put forward in the early 1990s and gradually applied in practice. A series of corresponding flow field calculations are completed by giving a series of different inlet passage boundaries and using three-dimensional turbulent flow numerical simulation method. According to the flow field calculation results, geometric parameters of the inlet passage are modified and optimized one by one, which becomes the basic idea for optimizing hydraulic design of the pumping station inlet passage.

8. 4. 4   Recommended dimensions of inlet passage

A large number of numerical calculations and model test data show that the hydraulic design criteria of inlet passage are closely related to the basic flow pattern in the channel, and the determination of each geometric parameter is subject to their influence on the flow pattern.

Based on the test and CFD optimized hydraulic calculation results, the recommended primary control dimensions for a one-sided inlet passage are listed in Table 8.1.

The height of impeller center is the most important parameter in flow passage design. The higher the height, the better the flow pattern at the pump inlet, but the more civil investment is required for the pumping station. This contradiction is especially pronounced in elbow-shaped inlet passages. The recommended values in Table 8.1 take into account both inlet flow pattern and civil investment requirements.

The length and width of the straight section of the flow passage have little influence on the flow field at the inlet of the pump impeller chamber. The length can generally be determined by the structural arrangement of the upper part of the pump house along the flow direction; the width of the flow passage can be determined by the requirements of the arrangement of the unit center distance, etc.

第 9 章 泵 房

泵房是安装主机组、辅助设备、电气设备、管路及其他设备的建筑物，是整个泵站工程的主体，它为机电设备、运行管理及维修人员提供必要的工作条件。因此，合理地设计泵房对节约工程投资，延长机电设备使用寿命，保证泵站安全高效运行有着重要意义。

泵房设计应遵循以下原则：

(1) 在满足设备安装、检修及安全运行的前提下，泵房尺寸和布置应尽量紧凑。

(2) 在各种工作条件下应满足稳定要求，各构件应满足强度和刚度要求，抗震性能良好。

(3) 充分满足通风、散热、采光、防潮、防火和低噪声等要求。

(4) 保证水下部分及输水结构不渗不漏。

(5) 应注意建筑造型，并与环境协调。

9.1 泵房的结构形式

9.1.1 泵房结构类型及演变过程

影响泵房结构形式的因素有：水泵机组的形式及容量；进出水位变幅；站址处的地基条件。

按泵房位置变动与否，将泵房分为固定式泵房和移动式泵房两大类。

(1) 移动式泵房又可分为泵船、泵车。泵船既可随水位变化做升降移动，又可做平面移动；泵车一般只固定在一处随水位变化做升降移动。泵船用于河网湖区，小而灵活机动；泵车用于水源水位变化幅度较大的地区，如从水库取水的泵站。

(2) 固定式泵房按其基础及水下结构的特点，可分为分基型泵房、干室型泵房、湿室型泵房和块基型泵房四类。

各种泵房的演变过程如下：

(1) 对于安装中小型卧式水泵的泵房，因单机流量小，有效吸程大，在水源水位变幅较小时，不需要水下结构，故采用最简单和经济的分基型泵房。

(2) 随着水泵口径和水源水位变幅的加大，一方面机组基础的单位面积重量增大，另一方面泵房要防止外水渗入，故需要将机组基础和泵房基础合建成一封闭的干室，从而形成了干室型泵房。

(3) 如果干室较深，则不利于通风、采光和防潮，同时要承受较大的浮托力和侧向压力，会使工程造价提高，这时采用立式机组可能比较合理。立式泵为了启动方便，将叶轮淹没于水下一定深度，这就使得进水池移至泵房下部，形成湿室型泵房。

(4) 当水泵流量很大时，要求进水流态更加均匀、对称，要用专门设计的进水流道来改善水泵的入流条件，同时，因机组尺寸及重量大，泵房的受力及结构复杂，对其整体性和稳定性要求较高，便将下部连同进水流道浇筑成大块整体基础，使之成为块基型泵房。

（5）当水源水位变幅很大时，各种固定式泵房都难以适应，出于技术和经济上的考虑，采用可随水位移动的浮船式或缆车式泵房会更加有利。

9.1.2 分基型泵房

分基型泵房是中小型泵站常采用的结构形式。它的主要特点如下：

（1）泵房的房屋基础与机组的基础分开。

（2）无水下部分，结构简单，施工方便。

（3）进水池与泵房分开，泵房高于进水池水面。

由于机组基础与房屋基础分开，因此，机组运行时的振动不至于影响整个泵房。泵房位于地面以上，通风、采光和防潮条件都较好，机组运行、检修方便。站前挡土墙采用直立式［图 9.1（a）］或斜坡［图 9.1（b）］两种。

分基型泵房适用于水源水位变幅较小、安装卧式机组的场合。若水源水位变幅较大，为了防止洪水位时泵房受淹，可在站前修建防洪闸进行水位调控。还可在泵房前岸坡上修建挡水墙，但须注意洪水位对地基的不利影响，防止地基渗水。

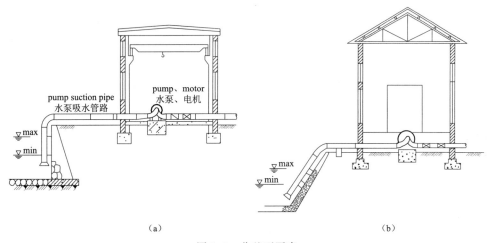

（a）   （b）

图 9.1 分基型泵房

Fig. 9.1 Separate-footing pump house

（a）直立挡土墙护坡；（b）斜坡

（a）Vertical retaining wall slope type；（b）Inclined retaining wall type

9.1.3 干室型泵房

当水源水位变幅超过一定范围时，若采用分基型泵房就不能满足低水位时水泵吸上高度的要求，在高水位时还易造成向泵房内渗水，影响泵站的安全和正常运行。为此，可将泵房底板适当降低并和侧墙用钢筋混凝土整体浇筑，形成一个不透水的泵室，这类泵房称为干室型泵房，如图 9.2 所示。

干室型泵房的平面形状常采用矩形和圆形两种，矩形泵房形式适用于多机组，圆形泵房适用于机组台数较少的情况。在水位变幅较大的情况下，泵房高度较大，为充分利用空间，往往建有楼板，将泵房分为上下两层，下层安装水泵、动力机、管道等，为水泵层；上层安装电气设备，为电气层。如采用立式机组，电动机与水泵以长轴相连，电动机安装在上层。

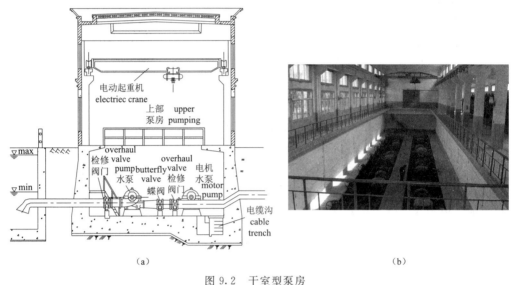

（a）　　　　　　　　　　　　　　　　　　　　（b）

图 9.2　干室型泵房
Fig. 9.2　Dry-pit pump house
（a）泵房剖面；（b）泵房内景
（a）Pump house profile；（b）Inside view of pump house

干室型泵房的底板也是机组的基础，采用钢筋混凝土浇筑，但应注意防渗。为排除无法避免的积水，底板表面要有一定坡度，并在最低处或四周设置集水沟和集水井或集水廊道，并设置排水泵。干室内通风不畅，应安装机械通风设施。

水泵进出水管上均应设置检修闸阀，以便高水位时检修水泵。

9.1.4　湿室型泵房

采用中小型立式轴流泵和导叶式混流泵时，可以将进水池设于泵房下部，形成一个具有自由水面的泵室，称为湿室型泵房，在我国平原、河网地区的低扬程泵站中应用最为广泛。这种泵房一般分为两层：上层为电机层，安装电动机和电气设备；下层为水泵层，安装水泵和吸水管。有时也采用封闭式有压进水池，则可分为三层：下层湿室，中层为水泵层，上层为电机层。

湿室型泵房的结构形式较多，可分为墩墙式、圆筒式、箱式、排架式、圬工泵房等，以下介绍常用的墩墙式、排架式和圬工泵房三种。

1. 墩墙式

墩墙式泵房（图9.3）适用于水源水位变幅在4～5m以下，地基条件较好的中小型立式半调节轴流泵站。它的主要特点是：

（1）水下部分除进水侧外，其他三面都有挡土墙。

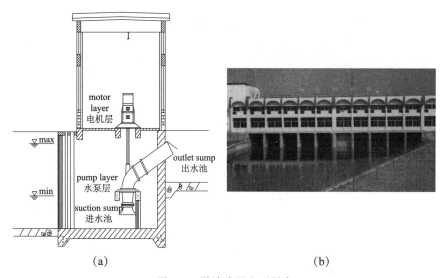

图 9.3 墩墙式湿室型泵房

Fig. 9.3 Wet-pit pump house with baffle-wall

(a) 泵房剖面；(b) 泵房外景

(a) Pump house profile；(b) Pump house exterion

(2) 水下结构由底板、隔墩和挡土墙围成。

(3) 相邻机组用隔墩分开，隔成多个进水池，每台机组有一单独集水室。

这种形式的优点是进水条件较好，各台机组可以单独检修，墩墙和底板可以采用浆砌块石等当地材料。主要缺点是墙后土压力很大，为了满足抗滑稳定要求，需增加泵房自重，因此，地基应力较大，要求地基有较高的承载能力。

2. 排架式

泵房下部为钢筋混凝土的排架，四周临水，用栈桥与岸上相通，出水管架设在桥上，也可沿岸坡架设。这种泵房结构较轻，钢筋混凝土结构少，地基应力小而均匀。由于没有侧向土压力，不必考虑抗滑稳定问题。缺点是护坡工程量大，如水位变幅较大时，机组传动轴较长，安装、检修及运行均不方便。排架式泵房（图 9.4）适用于安装中小型立式机组，水源水位变幅在 8~15m，地基条件较好的场合。

3. 圬工泵房

低洼地区以排涝为主的泵站，水位变化不大，扬程很低仅有 1~2m，水泵采用超低扬程的轴流泵，没有出水弯管，而以圬工出水室取代，故名圬工泵房。充分利用圬工结构，采用明池出水的圬工泵房分为三层，上层为电机层，中层为出水室，下层为封闭的进水室，如图 9.5 所示。

9.1.5 块基型泵房

大型泵站对泵房的稳定性和整体性有较高的要求，因此常将进水流道与泵房底板整体浇筑，形成块状基础结构作为整个泵房的基础，这种泵房称为块基型泵房，如图 9.6 所示。

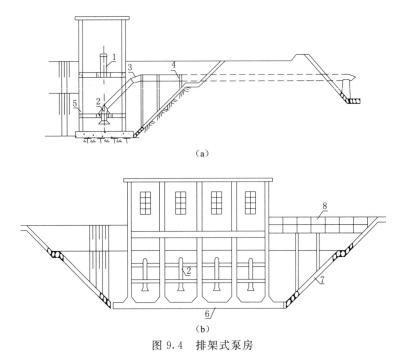

图 9.4 排架式泵房

Fig. 9.4 Wet-pit pump house with trestles bent

（a）剖面图；（b）立面图

（a）Profile；（b）Front view

1—电动机 electric motor；2—水泵 pump；3—出水管 outlet pipe；

4—穿堤涵管 culvert through embankment；5—排架 bent；

6—底板 floor；7—边坡 slope；8—栈桥 trestle

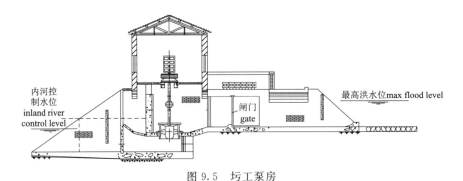

图 9.5 坞工泵房

Fig. 9.5 Masonry pump house

1. 泵房结构特点

块基型泵房整体性好，抗震能力强，适用立式、卧式和贯流式各种机组。其中以用于立式机组的泵房结构较为复杂，通常由下至上分为：进水流道层、水泵层、联轴层和电机层。进水流道层通常布置进水流道、廊道、空箱等，水泵层安装主水泵和供、排水设备，联轴层主要安装联轴器、电缆及油气水管路，电机层安装电动机和电气设备及其他辅助设备。

170

2. 泵房分类

按照泵房是否直接挡水和泵房与堤防的关系可分为：

(1) 堤身式。图9.6所示为立式和卧式两种堤身式块基型泵房。该类泵房出水流道与泵房整体浇筑，站身直接挡水，与左右侧堤防相连，直接承受上下游水位差产生的水平水压力及渗透压力。堤身式泵站的出水管路较长，适合于低扬程、大流量泵站及上下游水位差较小的场合。

(2) 堤后式。图9.7所示为堤后式块基型泵房。该类泵房通过堤防挡水防洪，泵房不直接承受上下游水位差产生的水平水压力及渗透压力。挡水建筑与站身分设，堤后式泵房适合于扬程较高、上下游水位差较大的场合。

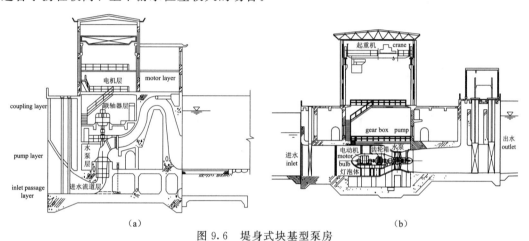

(a)　　　　　　　　　　　　　　　　(b)

图 9.6　堤身式块基型泵房

Fig. 9.6　Water retaining block-foundation pump house

(a) 立式；(b) 卧式

(a) Vertical；(b) Horizontal

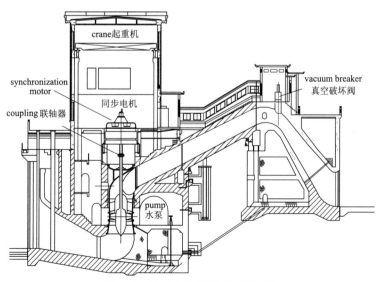

图 9.7　堤后式块基型泵房剖面图

Fig. 9.7　Profile of block-foundation pump house at levee-toe

9.2 泵房内部布置与主要尺寸确定

泵房内部布置,是对泵房内主机组、电气设备、辅助设备、管道、检修间、门窗及过道等统一安排,要在满足机组安装、检修、运行要求及泵房结构布置要求的前提下,尽量简单、紧凑、整齐、美观。

9.2.1 卧式机组泵房布置

1. 主机组布置

(1)一列式布置。各机组位于同一直线上,优点是布置整齐,可以缩小泵房跨度。卧式、立式泵站大多采用这个布置形式,如图 9.8(a)、(b)所示。

(2)双列交错式布置。当机组台数较多时,为了缩短泵房长度,可采用双列交错排列,但这种布置形式增加了泵房的跨度,如图 9.8(c)所示。

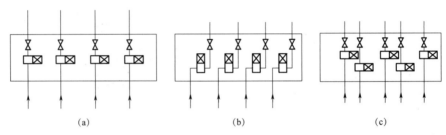

(a)　　　　　　　　　(b)　　　　　　　　　(c)

图 9.8 卧式机组布置示意图

Fig. 9.8 Horizontal unit layout

2. 配电设备布置

配电设备布置有集中布置和分散布置两种。分散布置是将配电柜布置在两台电动机中间靠墙的空地上,这时机房无须加宽。集中布置根据其在泵房中的位置,又分为两种型式,即一端式布置和一侧式布置。

(1)一端式布置。如图 9.9(a)所示,在泵房进线端建单独的配电间。其优点是机房跨度小,进出水侧都可以开窗,有利于通风及采光。缺点是增加了跨度,自然通风条件较差。

(2)一侧式布置。如图 9.9(b)所示,在泵房一侧布置配电柜。其优点是有利于监视机组的运行。为了不增加泵房跨度,可设副厂房布置配电柜。

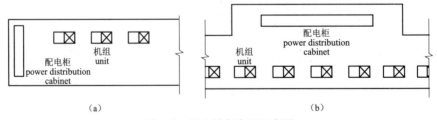

(a)　　　　　　　　　　　　　　(b)

图 9.9 配电设备布置示意图

Fig. 9.9 Schematic diagram of distribution facility

配电间的尺寸主要取决于配电柜的数目及其规格尺寸，以及操作维修空间。不靠墙安装的配电柜应有不小于 0.8m 宽的通道，以利于检修，柜前应留有 1.5～2.0m 的操作空间。

为保持配电间干燥，配电间的地板应略高出泵房地面。配电间一般都应单设一个外开的便门，以防事故之用。干室型泵房配电间地面高程应高于挡水墙外的最高水位，以防受潮和高水威胁。

3. 检修间布置

为装配及检修机组，在泵房靠近大门的一端设置检修间，其平面尺寸要求能够放下泵房内部的最大设备或部件，并便于拆卸检修，同时还要留有空地存放工具等杂物。对于不专设吊车的泵站，如机组容量较小或机组间距较大时，可原地进行检修而不单设检修间。

4. 交通道布置

为便于管理人员工作巡视、阀门启闭及物件搬运，在进水或出水侧设交通道。主要通道宽度不小于 1.5m，工作通道宽度不小于 1.0m。水泵单列布置的泵房，交通道多布置在出水侧，并高出泵房底板一定高度。双列布置的泵房，通常在两列机组间设置交通道。

5. 充水系统布置

水泵的充水系统包括充水设备及抽气干、支管。其布置以不影响主机组检修、不增加泵房面积、便于工作人员操作为原则。充水设备一般布置在主机组之间靠进水侧的空地上，抽气干管可与充水设备同侧，在高程上可沿机组基础的地面铺设，也可以支承在高于地面 2m 以上的空间，然后用抽气支管与每台水泵相连。

6. 排水系统布置

排水系统用以排除水泵水封渗滴水及管阀漏水等。底板或泵房地面应向下游有一定倾斜，泵房内设排水干、支沟。支沟一般沿机组基础布置，但应与电缆沟分开以免电缆受潮，废水沿支沟汇集到干沟中，然后可向墙外自流排出，或汇集至泵房端部的集水井中由排水泵抽排至进水池。通常前者适用于分基型泵房，后者适用于干室型泵房。

7. 通风布置

分基型泵房的通风主要是通过合理布置门窗实现风压或热压自然通风，干室型泵房需采用机械强迫通风。

8. 辅助设备布置

大型卧式机组的辅助设备较多，其布置可参考立式机组的辅助设备布置。

9.2.2 立式机组泵房布置

湿室型泵站布置较为简单，一般分为水泵层和电机层。大型立式机组泵房多采用块基型，平面上有分缝、立面上有分层问题，现以块基型泵房内部布置为例，某些原则也可供其他形式的泵房参考。

1. 电机层

（1）主机组。大型立式泵进水流道尺寸很大，机组只能一列式布置，机组间距受两个因素控制，一是电动机的电压和功率，二是进水流道的进水宽度。所确定的间距应满足安装、运行巡视及拆卸的要求。

（2）配电设备及主控制室。高压配电柜及励磁设备一般设于出水侧，靠墙布置，柜前留有足够的操作宽度。主控制室集中布置自动化屏、控制计算机，进行自动化控制操作和监视，其位置宜设于泵房一端。

（3）检修间。一般设在靠大门的一端，桥式吊车应能进行装卸作业，检修间应设检修孔，垂直吊放转子。还需设工具储藏间、休息室等。

（4）吊物孔。作为吊运下部设备的孔洞，可设在进水侧楼板中，但会增加泵房跨度；还可设在检修间底板上，如检修间下无空间，可设吊物井与水泵层相通。吊物孔的尺寸应保证下层设备最大部件通过。

（5）风道。采用机械直接排风的泵房，在电动机周围设有环形风道，经楼板下面，将热风排出泵房。

（6）电缆道。一般设于出水侧配电柜楼板之下、出水流道之上的空间。

（7）楼梯。只作为上下层人行交通之用，一般宽 0.8~1.0m，布置在泵房两端或进水侧墙边。

此外，电机层内还有部分辅助设备，其布置见后述。门窗布置同卧式机组，其他零星设备均布置在主机组的空隙中，力求整齐紧凑。电机层布置如图 9.10 所示。

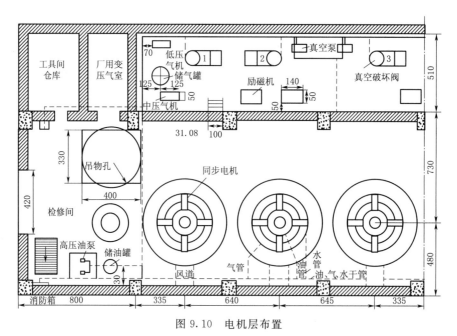

图 9.10　电机层布置

Fig. 9.10　Arrangement of motor layer

2. 水泵层

（1）主水泵。根据进水流道的要求，确定主泵的中心距，并与主机的中心调整一致，同时要考虑周围应留有安装拆卸的空间。若隔墩间距较小，不够装卸叶轮室外壳时，可将隔墩开孔，或将隔墩改为深梁，以扩大空间。

（2）过道。进出水侧均设过道，吊物孔所对的过道宽度，应能拖运水泵的最大部件。出水侧过道位于集水廊道盖板上，供、排水泵设在其中间或两端。

此外，进水流道顶板上应设进人孔，平时密封，孔径以能容人进出为宜。水泵层内设置的若干观察、测量仪表，可布置在适当位置。水泵层布置如图 9.11 所示。

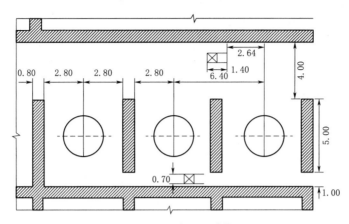

图 9.11　水泵层布置（单位：m）

Fig. 9.11　Arrangement of pump layer（unit：m）

3. 辅助设备布置

大型机组的辅助设备较多，有时在主泵房内布置不下，可在出水侧或两端设辅机房，集中布置相关设备。由图 9.10 可知，在电机层出水侧的辅机房内，布置了下列设备：

（1）励磁机。也可设于主泵房内，供同步电动机励磁用，改用可控硅励磁后，设备体积较小，可设于配电柜旁。

（2）真空泵。用于主机组及供、排水泵启动时抽真空。

（3）真空破坏阀。用于停机时虹吸管断流，设于虹吸出水流道顶部。

（4）中压空压机。向储气罐及储油箱供应压缩空气。

（5）低压空压机。用于需要使用压力为 $60\sim80N/cm^2$ 的设备作动力。

（6）储气罐。储存压力气体。

在电机层一端的辅机房或检修间内，还布置有高压油泵及储油箱。在主泵房或辅机房内还要布置压力油箱及油泵，并于楼板下铺设相应的油管。在主控室附近还要设有直流电源充电间。

在检修间或辅机房下层设有集油箱及油处理设备。其底板高程应使上下轴承油箱内的油能自流排入污油箱内，一般与联轴器检修间同高。

9.2.3　卧式机组泵房尺寸确定

1. 泵房跨度

泵房跨度应根据泵体的大小、进出水管道及其阀件的长度、安装检修及操作管理所必需的空间确定，并考虑进、出水侧所布置的走道宽度要求，其跨度也应与定型的屋架跨度或吊车跨度相适应。

为了避免漏气漏水以及便于拆装，泵房内外的进出水管道应采用金属管法兰连接。由于进出水管道的直径一般比水泵进口及出口的直径大，所以在水泵进口处需安装一个偏心渐缩接管，如图9.12中$b_缩$，出口处需安装一个同心渐扩接管$b_扩$。出水管道阀件的位置可根据具体情况确定。为了方便闸阀的拆卸，闸阀后往往接一伸缩节。此外，为了避免阀件重量传给水泵或其他设备，阀件下均设支墩进行支承。

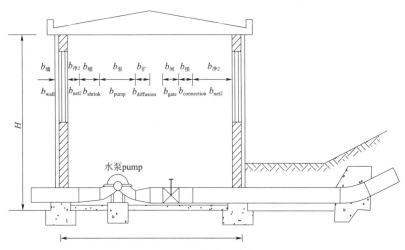

图 9.12 泵房跨度计算图

Fig. 9.12 Pump house span

2. 泵房长度

泵房长度主要根据机组及机组基础的长度，以及机组间距决定。机组间距根据机组大小、电机电压等级及操作维护等要求而定，可按表9.1选取。机组基础长L'，加上净空尺寸b为机组中心距L，L值应等于每台水泵要求的进水池宽度与池中隔墩厚度之和，如图9.13所示。机组中心距也就是泵房的柱距，在有配电间或检修间的泵房中，配电间或检修间的柱距可与机组间柱距相同，或根据设计需要确定。

表 9.1　　　　　　　　　　　泵房内部设备间距表

Table 9.1　　　　　　　　　Equipment spacing in pump house

流量 discharge/(m^3/s)	<0.5	0.5～1.5	>1.5
设备顶端与墙间的间距 spacing between top of the device and the wall/cm	70	100	120
设备与设备顶端的间距 distance between top of devices/cm	80～100	100～120	120～150
设备与墙间的间距 spacing between equipment and wall/cm	100	120	150
平行设备之间的间距 spacing between parallel devices/cm	100～120	120～150	150～200
高压或立式电动机组间的间距 spacing between high voltage or vertical motor units/cm	150	150～175	200

3. 泵房高度

图9.14中H即为泵房高度，它是指泵房地面与屋面大梁下缘之间的距离。分基型泵

房由于机组小、重量轻，故通常不专设吊车，此时泵房高度可根据水泵大小选取，一般不小于 3.5m。

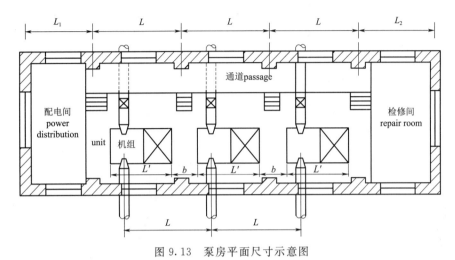

图 9.13 泵房平面尺寸示意图

Fig.9.13 Schematic diagram of plane dimension of pump house

设有吊车的泵房高度应满足吊车能从汽车的车厢中吊起最大设备，并能在已安装好的设备上空自由通行。泵房高度 H 可按式（9.1）计算：

$$H = h_1 + h_2 + h_3 + h_4 + h_5 + h_6 \tag{9.1}$$

式中：h_1 为车厢板离地面的高度；h_2 为垫块高，或吊起物底部与泵房室内地坪的距离，一般不小于 0.2m；h_3 为最高设备高度；h_4 为起重绳索的捆扎垂直长度，水泵为 $0.85b_0$，电动机为 $1.2b_0$，b_0 为水泵长度或电动机宽度；h_5 为吊钩极限高度；h_6 为单轨吊车梁高度，如采用桥式吊车，则为吊车高度与吊车顶至屋面大梁间的净空高度之和。

9.2.4 立式机组泵房尺寸确定

1. 湿室型泵房

湿室型泵房主机组一般为一列式布置，机组间距主要取决于下层进水池的进水要求。泵房宽度的确定与分基型泵房要求类似。墩墙式泵房通常在隔墩上留有检修门槽，以便起吊闸门，上部须设便桥，为此，泵房宽度加上必要的便桥宽度应与进水池长度对应。泵房各部高程的计算说明如下（图 9.15）。

（1）叶轮中心高程 $\nabla_{轮}$。该高程是泵房剖面设计中必须首先确定的高程，由水泵的汽蚀性能（$NSPH_r$）和最低运行水位确定。

（2）水泵吸水喇叭口高程 $\nabla_{进}$。取决于进口最低运行水位 $\nabla_{低}$ 及水泵尺寸，有

$$\nabla_{进} = \nabla_{低} - h_2 - h_3 \tag{9.2}$$

式中：h_2 为喇叭口至叶轮中心线高度；h_3 为水泵叶轮中心淹没深度。

（3）底板高程 $\nabla_{底}$。吸水喇叭口高程确定后，底板的高程决定于管口悬空高度 h_1，有

$$\nabla_{底} = \nabla_{进} - h_1 \tag{9.3}$$

（4）电机层楼板高程 $\nabla_{机}$。按进口最高水位 $\nabla_{高}$ 加上安全超高 δ 确定，有

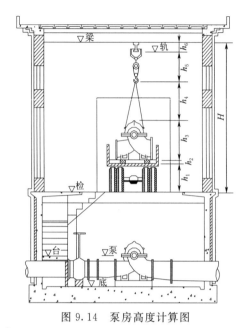

图 9.14　泵房高度计算图

Fig. 9.14　The height of pump house

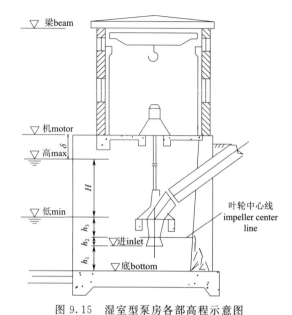

图 9.15　湿室型泵房各部高程示意图

Fig. 9.15　Elevations of wet-pit pump house

$$\nabla_{机} = \nabla_{高} + \delta \qquad (9.4)$$

δ 可以取 0.5～1.0m。

　　电机层楼板高程的确定还应与电动机和水泵连接所需要的中间轴的长度相对应，同时为防止地面雨水进入泵房，楼板应高于室外地面。

　　(5) 泵房屋面大梁下缘高程 $\nabla_{梁}$。屋面大梁的下缘至泵房楼板的垂直距离即为泵房的高度 H，其应满足起吊最大部件的要求。对于立式机组应考虑可以进行电动机转子抽芯和水泵的抽轴。

　　2. 块基型泵房

　　(1) 泵房各部高程的确定。图 9.16 为立式轴流泵块基型泵房各部位高程示意图，由图可见，整个泵房由下至上分为四层：进水流道层、水泵层、联轴层、电机层。

　　1) 水泵安装高程。由进口最低运行水位和水泵的汽蚀性能决定。

　　2) 进水流道底部高程 $\nabla_{底}$。根据进水流道的水力设计要求确定，叶轮中心高程减去进水流道高度。

　　3) 水泵层地面高程 $\nabla_{泵}$。根据水泵结构和检修拆装方便，确定泵坑高程 $\nabla_{坑}$，挡水前墙处流道顶板又能满足结构强度要求时，则可以 $\nabla_{坑}$ 作为 $\nabla_{泵}$。

　　4) 电机层地面高程 $\nabla_{机}$。根据联轴器位置高程和电动机轴伸长度确定。

　　(2) 泵房主要平面尺寸的确定。

　　1) 机组中心距。图 9.17 为机组间距示意图，由图可得，机组中心距为

$$L = L' + a \qquad (9.5)$$

式中：L' 为泵房站墩之间的净距，即进水流道的进口宽度；a 为中墩厚度，一般为 0.8～1.0m。对于泵房底板分块浇筑、中间有缝墩的机组，如缝墩厚度为 c（一般为 0.6～

0.8m)，则式中 $a=2c$。

2）泵房长度。底板长度（垂直水流向）由进水流道的宽度和隔墩、缝墩及边墩等结构尺寸决定；上层厂房长度在底板长度的基础上加上检修间长度。

3）泵房宽度。泵房宽度是指顺水流方向的宽度。下层底板宽度由进水流道长度、流道后廊道、空箱尺寸及出水流道布置等决定，空箱尺寸一般由泵房稳定要求决定；上层主泵房主要布置电动机、电气设备及油压装置等，考虑到泵房内大型物件的垂直运输及工作人员巡视，还需在电机层开设吊物孔和楼梯孔，主泵房宽度即由上述设备布置及交通要求而定，同时还需兼顾起吊设备的标准跨度。

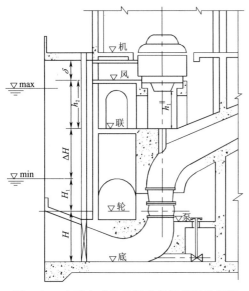

图 9.16 大型立式机组泵房各部高程示意图

Fig. 9.16 Elevations of pump house of large vertical unit

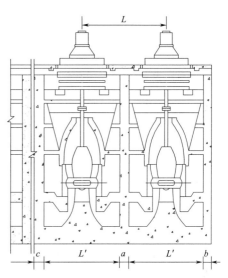

图 9.17 机组间距示意图

Fig. 9.17 Unit spacing

9.3 泵房整体稳定分析

初步拟定泵房尺寸后，必须进行整体稳定计算，泵房稳定分析包括：抗倾、抗滑、抗浮和地基稳定校核，还有地下轮廓线的设计计算等内容。稳定分析可取一个典型机组段作为计算单元，台数较少时，可直接取一块底板作为计算单元。要求泵房整体在外力和内部荷载的共同作用下，不发生倾覆、滑动或浮起等破坏。否则，必须根据计算结果对泵房布置和尺寸进行修改。

9.3.1 荷载组合与计算工况

1. 荷载组合

泵房在施工、运行、检修等不同时期所受的外部作用力、内部荷载也不同，必须选择最不利的情况进行计算校核。实践中，往往很难断定哪一种情况最危险，需要同时计算几种不同情况进行比较分析。荷载组合见表 9.2，表中"√""—"分别表示计及和不计及。

表 9.2 荷 载 组 合

Table 9.2 Load combination

荷载组合 load combination	计算工况 calculation conditions	荷 载 load							
		自重 self-weight	静水压力 hydrostatic pressure	扬压力 uplifting pressure	土压力 soil pressure	泥沙压力 sediment pressure	波浪压力 wave pressure	地震作用 earthquake action	其他荷载 other loads
基本组合 basic combination	完建期 completion period	√	—	—	√	—	—	—	√
	设计运用期 design usage period	√	√	√	√	√	√	—	√
特殊组合 special combination	施工期 construction period	√	—	—	√	—	—	—	√
	检修期 maintenance period	√	√	√	√	√	—	—	√
	校核运用期 verification operating period	√	√	√	√	√	√	—	—
	地震期 earthquake period	√	√	√	√	√	√	√	

2. 计算工况

泵房稳定计算通常需考虑以下几种计算工况情况。

（1）完建期。土建及安装工程已完成，但未拆除施工围堰，泵站前后无水。泵房主要承受建筑物自重及各种机电设备自重，还有侧向土压力、地下水压力等。此时所受的水平力较小，抗滑问题不大，但可能产生最大的地基应力。

（2）正常运行期。在设计运行时，泵房进出水侧均有水，除自重外，还承受水压力、土压力、底板上部的静水压力及下部的扬压力等。此时所受的水平力较大，除验算地基应力外，还要进行防滑稳定计算。

（3）检修及调相期。检修期一般在低水位时进行。视检修方式不同，有时抽空独立的进水池或进水流道进行逐台检修，有时需将前池、进水池水全部抽空。大型同步电动机有可能需要做调相运行，此时常将进水流道内水抽空，泵空转运行。

（4）校核情况。通常指泵站遭遇校核水位及地震、止水失效等非常情况。这些情况下的外荷载变化很大，为确保工程安全，需要进行校核验算，但地震力不与校核水位组合。

运行期泵房所受的荷载如图 9.18 所示。

9.3.2 泵房稳定计算

1. 抗滑稳定计算

对于修建在软土地基上的泵站，由于地基承载力较小，泵房在承受水平荷载和垂直荷

载的情况下，可能发生滑动破坏。滑动包括表面滑动和深层滑动，当发生表面滑动时，采用下式来进行抗滑稳定计算

$$K_c = \frac{f \sum V}{\sum H} \geqslant [K]$$ (9.6)

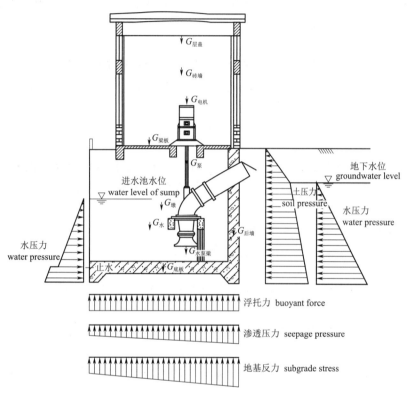

图 9.18　湿室型泵房运行期荷载

Fig. 9.18　Load of wet-pit pump house during operation period

当泵房基础前后均设有齿墙，而齿墙又较深时，泵房将连同齿墙间的土体滑动，此时的抗滑稳定计算公式为

$$K'_c = \frac{f_0 \sum V + CA}{\sum H} \geqslant [K]$$ (9.7)

式中：$\sum V$ 为垂直荷载（当有齿墙时，应包括齿墙间土体的重量）；$\sum H$ 为水平荷载；f 为底板与地基间摩擦系数；f_0 为沿滑动面土壤颗粒之间的摩擦系数，$f_0 = \tan\varphi$，φ 为土壤的内摩擦角；C 为齿墙间滑动面上土体的凝聚力，取室内试验值的 $1/5 \sim 1/3$；A 为齿墙间土体的剪切面积；K_c、K'_c 为抗滑稳定安全系数；$[K]$ 为抗滑稳定安全系数允许值，由建筑物的级别根据相关规范确定。

若计算出的 K_c 及 K'_c 值远大于 $[K]$，若非结构布置需要，说明泵房断面尺寸定得太大，可以考虑减小，以节约投资；若计算出的 K_c 及 K'_c 值小于 $[K]$，说明泵房不稳定，应采取措施，直至满足要求。调整 K_c 及 K'_c 值一般有如下措施：

（1）改变泵房结构尺寸或上部结构及设备布置，必要时改变泵房型式。

（2）加长防渗铺盖，延长渗径，以减小底板下的渗透压力。

（3）降低墙后填土高度或控制回填土料或控制地下水位，以减小水平推力。

（4）空箱内填土或填石，通过增加垂直荷重以加大抗滑力。

（5）设置钢筋混凝土阻滑板，以增加泵房抗滑稳定性。但需注意，在未加阻滑板时，泵房抗滑稳定安全系数必须在 1.0 以上。

2. 抗浮稳定计算

当泵房承受很大浮托力，有可能使泵房失稳时，应进行抗浮计算。一般计算情况为泵房刚建好，机组未安装，四周未填土，此时泵房四周达设计最高洪水位。抗浮稳定按下式计算

$$K_f = \frac{\sum V}{V_f} \geqslant [K_f] \tag{9.8}$$

式中：$\sum V$ 为全部垂直荷载；V_f 为扬压力。$[K_f]$ 为泵房抗浮安全系数的允许值，不分泵站级别和地基类别，基本荷载组合为 1.1，特殊荷载组合为 1.05。

3. 地基应力计算

泵房基础底面的地基应力按下式确定：

$$\sigma_{\min}^{\max} = \frac{\sum V}{F} \pm \frac{\sum M}{W} \tag{9.9}$$

式中：$\sum V$ 为全部垂直荷载；F 为基础底板面积；$\sum M$ 为全部荷载对底板中心的力矩和；W 为底板截面（底面）的抗弯模量。

计算出的最大地基应力必须小于地基容许应力，即地基容许承载力，最小地基应力必须大于零，即不出现拉应力。

为了不致产生过大的不均匀沉陷，地基应力的不均匀系数应在规范规定的范围内，即

$$K = \frac{\sigma_{\max}}{\sigma_{\min}} \leqslant [K] \tag{9.10}$$

当地基应力不满足要求时，通过调整局部布置或采取适当措施以达到稳定要求。

4. 地下轮廓线设计

与其他水工建筑物一样，防渗设计是泵站设计中的一项重要内容。渗透破坏通常有两种形式，黏土地基为流土，砂性地基为管涌。为了防止地基的渗透变形，除要求选择好泵房地下轮廓线外，还必须认真做好防渗和排水设施，以确保建筑物安全。

首先，可按勃莱或莱因法确定最小渗径长度 L，该值必须大于等于规定值，即

$$L \geqslant C \Delta H \tag{9.11}$$

式中：ΔH 为上下游最大水头差；C 为勃莱系数或莱因系数。

当实际渗径长度不足时，通常根据地基土质情况在上游增设防渗铺盖，增加水平渗径长度，或通过在底板下设齿墙、板桩增加垂直渗径长度。

Chapter 9　Pump House

Pump house is a major structure that installs main pumping unit, auxiliary equipment, electrical equipment, pipelines and other equipment, and is the main body of the whole pumping station project. It provides necessary working conditions for electromechanical equipment, operation management and maintenance personnel. Therefore, rational design of pump house is of great significance for saving project investment, prolonging the service life of electromechanical equipment and ensuring safe and efficient operation of pumping station.

The pump house design shall follow the following principles:

(1) On the premise of meeting the installation, maintenance and safe operation of the equipment, the size and layout of the pump house should be as compact as possible.

(2) The stability requirements should be met under various working conditions, and each component should meet the strength and stiffness requirements, with good anti-seismic performance.

(3) Fully meet the requirements of ventilation, heat dissipation, lighting, moisture-proof, fire protection and low noise.

(4) To ensure that the underwater part and the water delivery structure do not ooze and leak.

(5) Pay attention to architectural modeling and coordination with the environment.

9.1　Type of pump house

9.1.1　Type and Evolution Process of Pump House Structure

The coefficients affecting the structural form of pump house include: type and capacity of pump units; variation of inlet and outlet water level; foundation conditions of the station site.

According to whether the position of the pump house changes or not, the pump house is classified into two categories: stationary pump house and movable pump house.

(1) The movable pump house can be classified into pump boat and pump truck. Pump boat can move up and down with the change of water level, and can move in plane. Pump truck is usually fixed in one place to move up and down with the change of water level. Pump boats are used in river network and lake areas, which are small and flexible; pump trucks are used in areas with large changes in water level, such as pumping stations

that take water from reservoirs.

(2) According to the characteristics of its foundation and underwater structure, stationary pump house can be classified into four types: separate-footing, dry-pit, wet-pit and block-foundation.

The evolution process of various pump houses is as follows:

(1) For the pump house with medium and small horizontal pumps, the simplest and economical separate-footing pump house is adopted because of the small flow rate of single machine and large effective suction range, and when the variation of water level is small, the underwater structure is not needed.

(2) With the increase of pump diameter and water level variation, on the one hand, the unit area weight of unit foundation increases; on the other hand, to prevent the infiltration of external water in pump house, it is necessary to construct a closed dry house by combining unit foundation and pump house foundation, thus forming a dry house type pump house.

(3) If the dry-pit is deep, it is not conducive to ventilation, lighting and moisture control, while bearing a larger buoyancy force and lateral pressure, which will increase the cost of the project, which is more reasonable to use vertical units. In order to start up conveniently, the vertical pump submerges the impeller to a certain depth under water, which makes the sump move to the lower part of the pump house, forming a wet-pit pump house.

(4) When the flow rate of the pump is large, it is required that the flow pattern be more uniform and symmetrical. A specially designed inlet passage should be used to improve the inflow conditions of the pump. At the same time, because of the large size and weight of the unit, the complex force and structure of the pump house, and the high requirement for its integrity and stability, the lower part together with the inlet passage is poured into a large integral foundation, making it a block-foundation pump house.

(5) When the water level varies greatly, all kinds of stationary pump houses are difficult to adapt to. For technical and economic considerations, it would be more advantageous to adopt pump boat or pump vehicle that can move with the water level.

9. 1. 2   Separate-footing pump house

The separate-footing pump house is a common structural form used in medium and small pumping stations. Its main features are:

(1) The housing foundation of the pump house is separated from that of the unit.

(2) No underwater part, simple structure and convenient construction.

(3) The sump is separated from the pump house, which is higher than the water surface of the sump.

As the unit foundation is separated from the house foundation, the vibration during operation of the unit will not affect the whole pump house. Pump house is located above the ground, with good ventilation, lighting and moisture protection, and convenient operation and maintenance of the unit. The retaining wall in front of the station is vertical [see Fig. 9. 1 (a)] or inclined [see Fig. 9. 1 (b)] .

Separate-footing pump house is suitable for installing horizontal unit with small variation of water level. If the water level of water source varies greatly, in order to prevent the pump house from flooding when the flood level is high, a flood control gate can be built in front of the station to regulate the water level. The retaining wall can also be built on the front bank slope of the pump house, but the adverse effect of flood level on the subgrade must be noted to prevent the subgrade from seepage.

9. 1. 3　Dry-pit pump house

When the variation of water level exceeds a certain range, the requirement of suction height of pump can not be met if the separate-footing pump house is used at low water level. Water seepage into the pump house is also easy at high water level, which affects the safety and normal operation of the pumping station. To make full use of space, the pump house floor can be lowered appropriately and the side wall can be cast with reinforced concrete as a whole to form an impervious pump house, which is called dry-pit pump house, as shown in Fig. 9. 2.

The flat shape of dry-pit pump house is usually rectangular or circular. The rectangular pump house is suitable for multiple units, while the circular pump house is suitable for situations with few units. In the case of large water level variation, the pump house has a large height. In order to make use of space, floors are often built. Pump house is classified into upper and lower floors. Pumps, power motors, pipes are installed on the lower floor, which is the pump layer; electrical equipment is installed on the upper floor, which is the electrical layer. If a vertical unit is used, the electric motor is connected with the pump by a long shaft, and the electric motor is installed on the upper layer.

Floor of dry-pit pump house is also the subgrade of unit. Reinforced concrete is used for casting, but seepage prevention should be paid attention to. In order to eliminate the unavoidable water accumulation, the floor surface should have a certain slope, and drainage ditches, wells or galleries should be set at the lowest or around, and drainage pumps should be set. Mechanical ventilation facilities should be installed if ventilation in dry rooms is not smooth.

Inspection gate valves shall be provided on the inlet and outlet water pipes of pumps for inspection of pumps at high water levels.

9. 1. 4　Wet-pit pump house

When small and medium-sized vertical axial flow pump and guide vane mixed flow pump are used, the sump can be set at the lower part of the pump house to form a pump house with free water surface, which is called wet-pit pump house. It is most widely used in low-head pumping stations in plains and river networks in China. This pump house is generally divided into two layers: the upper layer is the motor layer, the installation of motor and electrical equipment; the lower layer is the pump layer, the installation of pumps and suction pipes. Sometimes closed pressurized sump is also used, which can be classified into three layers, well, pump layer and motor layer.

There are many types of wet-pit pump house, which can be classified into baffle-wall, cylindrical, box, trestle-framed, masonry pump house, etc.

1. Baffle-wall pump house

The baffle-wall pump house (see Fig. 9. 3) is suitable for small and medium-sized vertical semi-regulated axial flow pumping station with variable water level below 4 – 5m and good foundation condition. Its main characteristics are:

(1) There are retaining walls on all three sides of the underwater part except the suction side.

(2) Underwater structures are enclosed by floor, baffle and retaining wall.

(3) The adjacent units are separated by baffles and divided into multiple sumps, and each unit has a separate catchment chamber.

The advantages of baffle-wall pump house are that there are good inlet conditions, each unit can be repaired separately, and the baffle-wall and floor can be made of local materials such as masonry stones. The main disadvantage is that the earth pressure behind the wall is large. In order to meet the requirements of sliding resistance stability, it is necessary to increase the weight of the pump house. Therefore, the subgrade stress is large, which requires the subgrade to have a higher bearing capacity.

2. Trestle-framed pump house

The lower part of the pump house is a bent of reinforced concrete surrounded by water, which is connected with the shore by a trestle, and the outlet pipe is erected on the bridge or along the bank slope. This kind of pump house has lighter structure, less reinforced concrete structure and small and uniform subgrade stress. Since there is no lateral soil pressure, the problem of sliding resistance stability need not be considered. The disadvantage is that the revetment works are large. If the water level varies greatly, the transmission shaft of the unit is longer and the installation, maintenance and operation are inconvenient. The trestle-framed pump house (Fig. 9. 4) is suitable for installation of small and medium-sized vertical units, where the water level varies from 8m to 15m and the subgrade conditions are good.

3. Masonry pump house

The pumping stations in low-lying areas mainly drain waterlogging with small change in water level and low head of only about 1 – 2m. Pumps adopt ultra-low-head axial flow pumps without water outlet bend, and are replaced by masonry outlet chamber, which is called masonry pump house. Making full use of masonry structure, pump house with open outlet sump is classified into three layers, the upper layer is motor layer, the middle layer is open outlet sump, and the lower layer is closed sump, as shown in Fig. 9. 5.

9. 1. 5   Block-foundation pump house

Large pumping stations have high requirements for the stability and integrity of the pump house, so the inlet passage and floor of the pump house are often poured as a whole to form a block foundation structure as the basis of the whole pump house, which is called block-foundation pump house, as shown in Fig. 9. 6.

1. Structural characteristics of pump house

Block-foundation pump house has good integrity and strong seismic resistance, and is suitable for vertical, horizontal and tubular units. Among them, the pump house structure used for vertical units is more complex, which is usually divided from bottom to top into: inlet passage layer, pump layer, coupling layer and motor layer. The inlet passage layer is usually equipped with inlet passages, corridors, empty boxes, etc. The pump layer is equipped with main pumps and supply and drainage equipment, the coupling layer is mainly equipped with couplings, cables and oil and gas pipelines, and the motor layer is equipped with motors, electrical equipment and other auxiliary equipment.

2. Classification of pump house

According to whether the pump house directly retains water and the relationship between the pump house and the dike, it can be classified into:

(1) Dike body type. Fig. 9. 6 shows two dike block-foundation pump houses, vertical and horizontal. The outpassage of this kind of pump house is integrally poured with the pump house, the station body directly retains water, and is connected with the left and right embankments, which directly bears the horizontal water pressure and osmotic pressure caused by the water level difference between upstream and downstream. The outlet pipeline of dike-body pumping station is long, which is suitable for low-head and large-discharge pumping station and occasions with small water level difference between upstream and downstream.

(2) Post-embankment type. Fig. 9. 7 shows the block-foundation pump house behind the dike. This kind of pump house retains water and prevents flood through dike, and the pump house does not directly bear the horizontal water pressure and osmotic pressure caused by the water level difference between upstream and downstream. The water retaining building is separated from the station, and the pump house behind the dike is suitable for occasions with high head and large water level difference between upstream and downstream.

9. 2   Internal layout of pump house and determination of main dimensions

The internal layout of the pump house is a unified arrangement for the main unit, electrical equipment, auxiliary equipment, pipelines, repair rooms, doors, windows and corridors in the pump house. It should be as simple, compact, neat and beautiful as possible under the premise of meeting the requirements for the installation, overhaul, operation of the unit and the structural layout of the pump house.

9.2.1   Layout of pump house for horizontal units

1. Major pump room layout

(1) Single column arrangement. Each pump unit is arranged in a column on the same straight line. The advantage is that the arrangement is neat and the span of the pump house can be reduced. Horizontal and vertical pumping stations mostly adopt this layout form, as shown in Fig. 9.8 (a) (b).

(2) Double column interleaving arrangement. When the number interleaving of units is large, in order to shorten the length of the pump house, double column interleaving arrangement can be used, but this arrangement form increases the span of the pump house, as shown in Fig. 9.8 (c).

2. Switch gear layout

The layout of switch gear is arranged in two types: centralized layout and decentralized layout. Decentralized layout is to arrange the switch cabinet on the space near the wall between the two motors, at this time the room does not need to be widened. Centralized layout can be classified into two types according to its position in the pump house, namely one-end layout and one-side layout.

(1) One-end layout is shown in Fig. 9.9 (a), and a separate switch room is built at the inlet end of the pump house. The advantages are that the room span is small, and windows can be opened at the water inlet and outlet sides, which is conducive to ventilation and lighting. The disadvantage is increased span and poor natural ventilation conditions.

(2) One-side layout is shown in Fig. 9.9 (b), and the distribution cabinet is arranged on one side of the pump house. Its advantage is that it is beneficial to monitor the operation of the unit. In order not to increase the span of the pump house, the auxiliary plant can be set up to arrange the distribution cabinet.

The size of switch room mainly depends on the number of the distribution cabinet, their specifications and sizes, and the operation and maintenance space. switch cabinets installed without walls shall have channels not less than 0.8m wide to facilitate maintenance, and 1.5 – 2.0m operation space shall be reserved in front of cabinets.

To keep the switch room dry, the floor of the switch room should be slightly higher than the floor of the pump house. The switch room should generally have a single open door to prevent accidents. The floor elevation of switch room of dry-room pump house shall be higher than the maximum water level outside the retaining wall to prevent dampness and high water threat.

3. Arrangement of repair rooms

To assemble and repair the unit, the repair room is set at one end of the pump house close to the door, and its plane size requires that the largest equipment or components inside the pump house can be put down, which is convenient for disassembly and maintenance, while leaving room for the storage of tools and other debris. For pumping stations without special cranes, if the unit capacity is small or the unit spacing is large, the maintenance can be carried out in situ without the need for a single repair room.

4. Arrangement of access gallery

In order to facilitate management personnel work inspection, valve opening and closing and object handling, traffic lanes are set at the water inlet or water outlet side. The main channel is not less than 1. 5m, and the working channel is not less than 1. 0m. Pumps are arranged in a single row, and access gallerys are mostly arranged at the water outlet side, which is higher than a certain height of the pump house floor. Pump houses arranged in two columns usually have traffic lanes between two rows of units.

5. Arrangement of priming system

The priming system of the pump includes water filling equipment, air pumping dryer and branch pipe. Its layout is based on the principle of not affecting the maintenance of the host group, not increasing the area of the pump house and facilitating the operation of the staff. Filling equipment is generally arranged on the open space near the water inlet side between the main units. The pumping main pipe can be on the same side with the filling equipment. It can be laid along the ground of the unit foundation at elevation, or can be supported in the space above 2m above the ground. Then it can be connected with each pump with the pumping branch pipe.

6. Arrangement of drainage system

Drainage system is used to discharge waste water from pump seal, cooling and leaky water from pipes and valves, etc. The floor or pump house ground shall be inclined downstream to a certain extent, and drainage trunk and branch ditch shall be set in the pump house. Branch ditches are generally arranged along the unit foundation, but they should be separated from cable ditches to avoid dampness of cables. Waste water is collected into dry ditches along the branch ditches, and then can be discharged out of the wall spontaneously or collected into the catchment wells at the end of the pump house and pumped to the sump by the drainage pump. Usually the former is suitable for the separate-footing pump house, while the latter is suitable for the dry-pit pump house.

7. Ventilation layout

Ventilation of separate-footing pump house is mainly realized by natural ventilation of air pressure or hot pressure through reasonable arrangement of doors and windows, and mechanical forced ventilation is required for dry-pit pump house.

8. Auxiliary equipment layout

There are many auxiliary equipments for large horizontal units, and their layout can be referred to that of vertical units.

9.2.2   Arrangement of vertical units

The wet-pit pump house is arranged simply and generally divided into pump layer and motor layer. The block-foundation pump houses for large vertical units are mostly used, with cracks in the plane and stratification problems in the elevation. The internal arrangement of block-foundation pump houses is described below. Some principles can also be used for reference of the pump houses of other type.

1. Motor layer

(1) Main unit. The inlet passage of large vertical pump of main unit is large in size, and the unit can only be arranged in one column. The unit spacing is controlled by two co-efficient, one is the voltage and power of the motor, the other is the inlet width of the inlet passage. The spacing determined shall meet the requirements for installation, operation inspection and disassembly.

(2) Distribution equipment and main control room. High-voltage distribution cabinet and excitation equipment are generally set on the outlet side, arranged against the wall, leaving sufficient operating width in front of the cabinet. The automatic screen and control computer are arranged centrally in the main control room for automatic control operation and monitoring, and their position should be set at one end of the pump house.

(3) Repair room. The repair room is usually located at the end near the door. The bridge crane should be able to handle the work. The repair room should be provided with overhaul holes and the rotor should be lifted vertically. Tool storage room, lounge, etc. are also required.

(4) Hoist-way. As a hole for lifting lower equipment, the hoist-way can be set in the floor of the intake side, but it will increase the span of the pump house. It can also be set on the floor of the repair room. If there is no space under the repair room, the hoist-way can be set to connect with the pump floor. The size of the hoist-way shall ensure that the largest part of the lower equipment passes through.

(5) Air duct. The air duct is a pump house with mechanical direct air exhaust. A ring-shaped air duct is arranged around the electric motor. Hot air is discharged from the pump house through the floor below.

(6) Cable duct. The cable ducts are generally located under the floor of the switch cabinet on the outlet side and above the outlet passage.

(7) Stairs. The stairs, generally 0.8 – 1.0m wide, are only used for pedestrian traffic on the upper and lower floors. They are arranged at both ends of the pump house or beside the intake side wall.

There are also many auxiliary equipment in the motor layer of other equipment, whose arrangement is described later. The doors and windows are arranged in the same horizontal unit, while other sporadic equipment are arranged in the space of the main unit in an effort to be neat and compact. The arrangement of motor layer is shown in Fig. 9.10.

2. Pump layer

(1) Main pump. The center distance of the main pump shall be determined according to the requirements of the inlet passage, which is consistent with the adjustment of the center of the main machine, and the space for installation and disassembly shall be reserved around it. If the spacing between the division pier is small enough to load and unload the impeller chamber, a hole shall be open at the division pier or the division pier can be changed to a deep beam to expand the space.

(2) Corridor. The inlet and outlet sides of the pumping station are all provided with corridors. The width of the corridor opposite the hoist-way shall be able to tow the largest part of the pump. The outlet side corridor is located on the cover plate of the catchment corridor, and the supply and drainage pumps are located in the middle or both ends of it.

Manholes shall be installed on the roof of other inlet passages, and shall be sealed at ordinary times. The pore size shall be suitable for accommodating people to enter and exit. Several observation and measurement instruments set in the pump layer can be arranged in appropriate positions. The arrangement of pump layer is shown in Fig. 9.11.

3. Arrangement of auxiliary equipment

There are many auxiliary equipments for large-scale units, sometimes they cannot be arranged in the main pump house. Auxiliary room can be set at the outlet side or both ends, and relevant equipment can be centralized. The following equipment is arranged in the auxiliary room at the outlet side of the motor layer, which is shown in Fig. 9.10:

(1) Exciter. The exciter can also be set in the main pump house for the excitation of synchronous motors. After switching to thyristor excitation, the equipment volume is small and can be set next to the switch cabinet.

(2) Vacuum pump. The vacuum pump used for vacuum extraction when the main unit and the supply and drainage pumps start.

(3) Vacuum breaking valve. Vacuum breaking valve used to shut off the siphon during shutdown, located at the top of the siphon outlet passage.

(4) Medium pressure air compressor. Supply compressed air to the gas tank and oil tank.

(5) Low pressure air compressor. Low pressure air compressor used to power equipment with a pressure of $60-80\text{N/cm}^2$.

(6) Storage tank. Storage of pressure gas.

In the auxiliary room or maintenance room at one end of the motor layer, high-pressure oil pumps and storage tanks are also arranged. Pressure tank and oil pump shall also be arranged in the main pump house or auxiliary machine room, and corresponding oil pipes shall be laid under the floor. There is also a charging room for DC power supply near the main control room.

Oil collecting tank and oil treatment equipment are provided in the maintenance room or the lower layer of auxiliary engine room. The floor elevation shall allow the oil in the upper and lower bearing oil tanks to flow into the dirty oil tank at the same height as that in the coupling repair room.

9. 2. 3   Dimensional determination of pump house for horizontal units

1. Span of pump house

The span of the pump house shall be determined according to the size of the pump body, the length of the inlet and outlet water pipes and their valves, the necessary space for installation, overhaul and operation management, and the width requirements of the walkways arranged on the inlet and outlet sides shall be considered. The span of the pump house shall also be compatible with the type roof frame span or crane span.

In order to avoid air leakage and water leakage and facilitate disassembly, the inlet and outlet water pipes inside and outside the pump house shall be connected by metal pipe flanges. Since the diameter of the inlet and outlet pipes is generally larger than that of the inlet and outlet of the pump, an eccentric tapering nozzle should be installed at the inlet of the pump. As shown in Fig. 9. 12, a concentric tapering nozzle should be installed at the outlet. The position of valve parts in the outlet pipe can be determined according to the specific conditions. In order to facilitate the disassembly of the gate valve, a telescopic joint is often connected behind the gate valve. In addition, to avoid the weight of the valve parts being passed to the pump or other equipment, baffles are provided under the valve parts for support.

2. Length of pump house

The length of pump house is mainly determined by the length of unit and unit foundation, as well as the unit spacing. Unit spacing is determined according to unit size, motor voltage level, operation and maintenance requirements, and can be selected according to Table 9. 1. The unit foundation length L', plus the clearance size b as the unit center distance L, the L value should be equal to the sum of the width of the sump required by each pump and the upsetting thickness of the septum in the sump, as shown in Fig. 9. 13. The center distance of the unit is also the column distance of the pump house. In the pump house with power distribution room or repair room, the column distance between the power distribution room or repair room can be the same as that between the units or determined according to the design needs.

3. Height of pump house

H in Fig. 9. 14 is the height of the pump house, which refers to the distance between the ground of the pump house and the lower edge of the roof girder. Because of the small unit size and light weight, no special crane is usually set in the separate-footing pump house. At this time, the height of the pump house can be selected according to the size of the pump, which is generally not less than 3.5m.

The height of the pump house equipped with a crane shall satisfy that the crane can lift the maximum equipment from the car compartment and can freely pass over the installed equipment. The pump house height H can be calculated by Eq. (9.1).

$$H = h_1 + h_2 + h_3 + h_4 + h_5 + h_6 \tag{9.1}$$

Where, h_1 is the height of the carriage board from the ground; h_2 is the height of the cushion block or the distance between the bottom of the lifting object and the floor of the pump house, which is generally not less than 0.2m; h_3 is the highest equipment height; h_4 is the bundling vertical length of the lifting rope, the pump is 0.85 b_0, the motor is 1.2 b_0, and b_0 is the pump length or motor width; h_5 is the limit height of the hook; h_6 is the height of the monorail crane girder. If a bridge crane is used, it is the sum of the height of the crane and the clearance height from the crane roof to the roof girder.

9.2.4 Dimensions of pump house of vertical unit

1. Wet-pit pump house

The main unit group of wet-pit pump house is generally arranged in a column, and the unit spacing mainly depends on the water intake requirements of the lower sump. The determination of pump house width is similar to the requirement of separate-footing pump house. The baffle-wall pump house usually has overhaul gate groove on the division pier to lift the gate, and the upper part must be provided with a temporary bridge. Therefore, the width of the pump house plus the necessary width of the temporary bridge should correspond to the length of the sump. The elevation calculation of each part of the pump house is illustrated as follows (see Fig. 9.15).

(1) The impeller center elevation $\nabla_{轮}$ is the elevation that must be determined first in the profile design of the pump house, which is determined by the cavitation performance ($NPSH_r$) of the pump and the lowest operating water level.

(2) The pump suction bellmouth elevation $\nabla_{进}$ depends on inlet minimum operating water level and pump size:

$$\nabla_{进} = \nabla_{低} - h_2 - h_3 \tag{9.2}$$

Where, h_2 is the height from the bellmouth to the center line of the impeller; h_3 is the submerged depth in the center of the impeller of the pump.

(3) After the elevation of the suction bellmouth is determined, the elevation of the floor $\nabla_{底}$ depends on the suspension height of the bellmouth h_1:

$$\nabla_{底} = \nabla_{进} - h_1 \tag{9.3}$$

(4) Floor elevation of motor floor is determined according to the maximum water level height of inlet plus safe superelevation:

$$\nabla_{机} = \nabla_{高} + \delta \tag{9.4}$$

δ can be taken from 0.5 to 1.0m.

The elevation of the floor of the motor layer should also correspond to the length of the intermediate shaft required for the connection of the motor and the pump. At the same time, in order to prevent the ground rainwater from entering the pump house, the floor should be higher than the outdoor floor.

(5) Elevation of lower edge of pump house girder $\nabla_{梁}$. The vertical distance from the lower edge of beam roof girder to the floor of pump house is the height H of pump house, and its height should meet the requirement of maximum lifting component. For vertical units, motor rotor core extraction and pump shaft extraction should be considered.

2. Block-foundation pump house

(1) Determination of elevation of each part of pump house. Fig. 9. 16 shows the elevation diagram of each part of the basic pump house of vertical axial flow pump block. It can be seen from the diagram that the whole pump house is classified into four layers from bottom to top: inlet passage layer, pump layer, coupling layer and motor layer.

1) The installation elevation of the pump is determined by the lowest operating water level at the inlet and the cavitation performance of the pump.

2) The bottom elevation of the inlet passage $\nabla_{底}$ is determined according to the hydraulic design requirements of the inlet passage, and the height of the impeller center is subtracted from the height of the inlet passage.

3) The ground elevation of the pump layer can be determined according to the structure of the pump and the convenience of maintenance, disassembly and installation. When the top plate of the passage at the water retaining front wall can meet the structural strength requirements, $\nabla_{坑}$ can be used as $\nabla_{泵}$.

4) The ground elevation of the motor layer is determined according to the position elevation of the coupling and the extension length of the motor shaft.

(2) Main plane dimensions of pump house.

1) Unit center distance. Fig. 9. 17 shows a schematic diagram of unit spacing. It can be obtained from the diagram that the unit center distance is

$$L = L' + a \qquad (9.5)$$

Where, L' is the net distance between the piers of the pump house, i. e. the width of the inlet passage; a is the thickness of the middle pier, which is generally 0. 8 – 1. 0m. For the unit with joint pier in the middle and block casting of pump house floor, if the thickness of joint pier is c (generally 0. 6 – 0. 8m), then $a = 2c$ in the formula.

2) Length of pump house. The length of the floor of the pump house (vertical water direction) is determined by the width of the inlet passage and the structural dimensions of the middle piers, joint pier and side pier; the length of the upper workshop is added with the length of the repair room on the basis of the length of the floor.

3) Width of pump house. Pump house width refers to the width along the flow direction. The width of the lower floor is determined by the length of the inlet passage, the

rear corridor of the passage, the size of the empty box and the layout of the outlet passage. The size of the empty box is generally determined by the stability requirements of the pump house. The upper main pump house is mainly equipped with motors, electrical equipment and oil pressure devices, etc. Considering the vertical transportation of large items in the pump house and the inspection of the staff, it is also necessary to open lifting and staircase holes in the motor layer. The width of the main pump house is determined by the above equipment layout and traffic requirements, and the standard span of the lifting equipment should also be taken into account.

9.3   Overall stability analysis of pump house

After preliminary drawing up the size of the pump house, the integral stability calculation must be carried out. The stability analysis of the pump house includes: tilt resistance, sliding resistance, buoyant resistance and subgrade stability check, as well as the design and calculation of the underground configuration line. For stability analysis, a typical unit section can be taken as the calculation unit, and when the number of units is small, a floor can be taken directly as the calculation unit. It is required that under the combined action of external force and internal load, the pump house as a whole should not suffer from overturning, sliding or floating damage. Otherwise, the layout and size of the pump house must be modified according to the calculation results.

9.3.1   Load combination and calculation conditions

1. Load combination

Pump house suffers different external forces and internal loads in different periods of construction, operation and maintenance, so the most unfavorable conditions must be selected for calculation and verification. In practice, it is often difficult to determine which situation is the most dangerous, and several different situations need to be calculated at the same time for comparative analysis. Load combinations are shown in Table 9.2, and "√" and "—" in the table indicate whether they are taken into account.

2. Calculation conditions

The following calculation conditions are usually considered in stability calculation of pump house.

(1) Civil and installation works have been completed during the completion period, but the construction cofferdam has not been removed and there is no water in front of and behind the pumping station. The pump house mainly bears the self-weight of buildings and various electromechanical equipment, as well as the soil pressure and groundwater pressure. At this time, the horizontal force is small and the problem of sliding resistance is not big, but the maximum subgrade stress may be generated.

(2) During normal operation period, when the pump house is designed and operated,

there is water on the inlet and outlet sides. Besides self-weight, it also bears water pressure, soil pressure, hydrostatic pressure at the upper part of the floor and uplift pressure at the lower part. In addition to checking the subgrade stress, the calculation of sliding resistance stability should also be carried out.

(3) Maintenance period and phase adjustment period are usually carried out at low water level. Depending on the mode of maintenance, sometimes the independent sump or inlet passage is emptied for one-by-one maintenance, and sometimes the forebay and sump are completely emptied. Large synchronous motors may need to be operated in phase adjustment, at which time the water in the inlet passage is often evacuated and the pump runs idly.

(4) Verification usually refers to the pumping station encountering abnormal conditions such as check water level, earthquake, watertight failure, etc. In these cases, the external loads vary greatly and checking calculations are required to ensure the safety of the project. However, the seismic force is not combined with check water level.

Loads on the pump house during operation are shown in Fig. 9.18.

9.3.2　Stability calculation of pump house

1. Sliding resistance stability calculation

For pumping stations built on weak subgrade, sliding occur when the pump house is subjected to horizontal and vertical loads due to the small bearing capacity of the subgrade. Sliding includes surface sliding and deep sliding. When surface sliding occurs, the following formula is used to calculate the sliding resistance stability

$$K_c = \frac{f \sum V}{\sum H} \geqslant [K] \tag{9.6}$$

When the subgrade of the pump house is provided with a toothed wall in front and back, and the toothed wall is deep, the pump house will slide along with the soil between the toothed walls. At this time, the formula for calculating the sliding resistance stability is as follows

$$K'_c = \frac{f_0 \sum V + CA}{\sum H} \geqslant [K] \tag{9.7}$$

where, $\sum V$ is vertical load (when there is a toothed wall, the weight of soil between the toothed walls should be included); $\sum H$ is horizontal load; f is friction coefficient between the floor and the subgrade; f_0 is friction coefficient between soil particles along the sliding surface, $f_0 = \tan\varphi$, and φ is the internal friction angle of soil; C is cohesion of soil on sliding surface between teeth walls, taking $1/5 - 1/3$ of laboratory test value; A is shear area of soil mass between tooth walls; K_c, K'_c is safety coefficient sliding resistance; $[K]$ is permissible value of safety coefficient of sliding resistance stability, determined by the level of building according to relevant specifications.

If the K_c and K'_c values are greater than $[K]$, and not required for the structural layout, it means that the section size of the pump house is set too large, it can be considered to reduce to save investment; if the K_c and K'_c values are less than $[K]$, it means that the pump house is unstable, measures should be taken until the requirements are met. There are generally the following measures to adjust K_c and K'_c:

(1) Change the size of pump house structure or upper structure and equipment layout, and change the pump house type if necessary.

(2) The anti-seepage cover is lengthened and the seepage path is extended to reduce the seepage pressure under the floor.

(3) To reduce the horizontal force, reduce the filling height behind the wall or control the backfill material or control the groundwater level.

(4) Fill the empty box with soil or rock, and increase the sliding resistance force by increasing the vertical load.

(5) Reinforced concrete sliding plate is set to increase the stability of pump house against sliding. However, it should be noted that the sliding resistance stability safety coefficient of the pump house must be above 1.0 when no sliding resistance plate is added.

2. Buoyant resistance stability calculation

When the pump house bears a large buoyancy, which may destabilize the pump house, anti-buoyancy calculation should be carried out. The general calculation is that the pump house has just been built, the unit has not been installed, and there is no soil filled around. At this time, the pump house is surrounded by the highest designed flood level. The buoyant resistance stability is calculated by the following formula

$$K_f = \frac{\Sigma V}{V_f} \geqslant [K_f] \tag{9.8}$$

Where, ΣV is total vertical load; V_f is uplift pressure. $[K_f]$ is the permissible value of buoyant resistance safety coefficient of pump house, regardless of pumping station rank and subgrade category, the basic load combination is 1.1, and the special load combination is 1.05.

3. Subgrade stress calculation

The stress on the bottom of pump house foundation is determined according to the following formula

$$\sigma_{min}^{max} = \frac{\Sigma V}{F} \pm \frac{\Sigma M}{W} \tag{9.9}$$

Where, ΣV is total vertical load; F is area of the base floor; ΣM is sum of the moments of the whole load on the center of the floor; W is flexural modulus of the floor section (bottom).

The calculated maximum subgrade stress must be less than the permissible stress of the subgrade, that is, the permissible bearing capacity of the subgrade, and the minimum subgrade stress must be greater than zero, that is, no tensile stress occurs.

In order not to cause excessive non-uniform settlement, the non-uniform coefficient of subgrade stress should be within the range specified in the code, that is

$$K = \frac{\sigma_{max}}{\sigma_{min}} \leqslant [K] \qquad (9.10)$$

When the subgrade stress does not meet the requirements, the stability requirements can be achieved by adjusting the local layout or applying appropriate measures.

4. Design of underground configuration line

Like other hydraulic structures, seepage control design is an important part of pumping station design. There are usually two forms of seepage failure: flowing soil for the clay foundation and piping for the sandy foundation. In order to prevent the seepage deformation of the subgrade, besides choosing the underground contour of the pump house, it is also necessary to do a good job of seepage prevention and drainage facilities to ensure the safety of the building.

Firstly, the minimum seepage length L can be determined by the Bouley's or Rein's method, which must be greater than or equal to the specified value, i. e.

$$L \geqslant C \Delta H \qquad (9.11)$$

Where, ΔH is maximum head difference between upstream and downstream; C is Bouley coefficient or Rein coefficient.

When the actual seepage path is insufficient, seepage-proof paving is usually added in the upstream to increase the horizontal seepage length according to the subgrade soil condition, or vertical seepage length is increased by setting toothed walls and sheet piles under the floor.

第10章 出水建筑物

泵站出水建筑物主要有出水池（或压力水箱）、出水流道、出水管道、出水渠道及断流设施等。它必须满足以下要求：

（1）保证有足够的稳定性和安全性。

（2）良好的出水流态，最大限度地回收水泵出口的水流动能。

（3）各部分接头及密封性能良好。

（4）在泄水和断流时应具有排气和破坏真空设施。

（5）便于施工和运行管理。

（6）节省土建投资。

10.1 出 水 池

10.1.1 出水池的作用与设计要求

出水池是连接出水管道和出水渠道的衔接建筑物，如图10.1所示。其主要作用是：

（1）汇集出水管道的来流，有时也起分流作用，向连接于出水池的几条干渠分流。

（2）防冲稳流，扩散出水管水流，将水流平顺地引入干渠，以免造成渠道冲刷。

（3）便于设置防止停泵时水流倒流的设施；便于设置检修和断流设施。

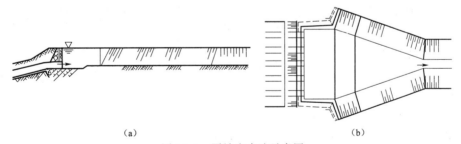

（a）　　　　　　　　　　　　　　　　（b）

图10.1　泵站出水池示意图

Fig. 10.1　Outlet sump of pumping station

（a）剖面图；（b）平面图

（a）Section；（b）Plane

出水池的位置一般高于泵房，一旦发生滑塌事故必将危及整个泵站的安全，因此，要求其结构上必须可靠稳定，一般应修筑在挖方上。出水池的防渗计算、地基应力校核、地基沉陷量计算等必须满足规范要求。出水池的高度必须保证最大流量时不发生漫溢，确保稳固可靠。出水池中水流平顺稳定，流速一般不超过2m/s。为防渠道被冲刷，干渠进口应有一定的护砌长度。

10.1.2　出水池类型

出水池的类型一般按水流方向的不同分为正向出水池、侧向出水池和多向分流式出水池三种类型。

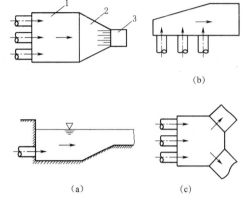

图 10.2　三种类型出水池

Fig. 10.2　Three types of outlet sump

(a) 正向出水池；(b) 侧向出水池；

(c) 多向分流式出水池

(a) Forward outlet sump；(b) Side outlet sump；

(c) Multi-directional diversion

1—出水池 outlet sump；2—过渡段 transition section；

3—干渠 trunk channel

(1) 正向出水池的管口水流方向与干渠水流方向一致［图 10.2 (a)］，水流顺直、水力性能好，应用较多。

(2) 侧向出水池的管口水流方向与干渠水流方向正交［图 10.2 (b)］，由于水流 90°转弯，水流交叉且与池壁相撞，水力性能较差，一般只有在正向出水无法布置的情况下才采用。

(3) 多向分流式出水池用于同时向多条干渠输水的情况，池中流态介于正向出水和侧向出水之间［图 10.2 (c)］，通过采取适当的导流措施，可以改善水力性能。

按出水池与泵房的关系，可分为分建式和合建式两种形式的出水池，前者将出水池与泵房分开建造，多用于高扬程泵站；后者将出水池与泵房建成一个整体，多用于低扬程泵站。

按断流方式的不同，又可分为拍门式、闸门式、虹吸管式、溢流堰式、自由出流式等不同形式的出水池。

按是否有自由水面，又可分为开敞式出水池和压力水箱两种类型，前者即通常简称的出水池，池内具有自由水面；后者则为封闭的有压管道。

10.1.3　出水池流态

图 10.3 为正向出水池的流态示意图。水流从出水管进入出水池后在立面方向和平面方向同时扩散，呈有限空间三维扩散淹没射流状态。在立面方向，在主流上部有大的漩滚区，下部有小范围漩滚；在平面方向，在出水管的两侧形成回流区。

由于漩滚区的存在，池内水流紊乱，若处理不当，将导致水力损失增加和出水池及渠道的冲刷。漩滚区的形状和大小（包括图中的立面扩散角 α、平面扩散角 β 及漩滚长度 L 等）与出水管的流态、管口淹没深度及出水池的几何边界有密切关系。

图 10.4 为侧向出水池的流态示意图。池内水流受到正面壁面的阻挡而形成反向回流，出流不畅，致使水面壅高、水力损失增加。壁面距管口越近，出水流态所受影响越大。

10.1.4　出水池尺寸的确定

图 10.5 所示为一淹没出流正向出水池，出水管以水平方向进入出水池，图中给出了出水池各几何参数的意义。

1. 出水管出口直径 D_0

为降低出口流速，使出水池中不产生水跃并减少出口损失，出水管出口直径宜取得大一些；另外，出水管出口直径也不宜过大，以免过分增大配套的拍门及有关尺寸。一般按出水管管口的平均流速在 $1.5\sim2.5$m/s 范围内选取。

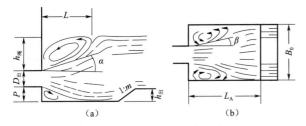

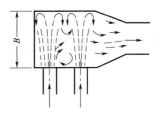

图 10.3　正向出水池内流态

Fig. 10.3　Flow pattern in front outlet sump

（a）立面；（b）平面

（a）Vertical section；（b）Plane section

图 10.4　侧向出水池内流态

Fig. 10.4　Flow pattern in side outlet sump

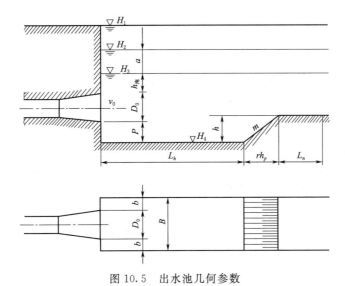

图 10.5　出水池几何参数

Fig. 10.5　Geometric parameters of the outlet sump

2. 淹深 $h_{淹}$

出水管管口应留有一定的淹没深度，其目的是不使出水管水流冲出水面、增加水力损失和水面漩滚。该淹没深度的取值与管口流速有关，一般取

$$h_{淹}=(2\sim3)\frac{v_0^2}{2g}(\text{m}) \tag{10.1}$$

3. 池底至管口下缘距离 P

为便于出水管道和拍门的安装，避免泥沙或杂物堵塞管口，出水管管口与出水池池底之间应留有一定的空间，一般取 $P>0.3$m。

4. 出水池墙顶高程和池底高程

出水池的高度应保证在最高水位时不发生漫溢。出水池墙顶高程按下式计算

$$\nabla_{池顶} = \nabla_{max} + h_{超高} \quad (m) \tag{10.2}$$

式中：$h_{超高}$ 为安全超高，与泵站流量有关，可参考表 10.1 选取；∇_{max} 为出水池最高水位，m。

表 10.1 出 水 池 的 安 全 超 高
Table 10.1 Safety superelevation of outlet sump

泵站流量 discharge/（m^3/s）	安全超高 $h_{超高}$ safety superelevation/m
<1	0.4
$1 \sim 6$	0.5
>6	0.6

出水池池底高程为

$$\nabla_{池底} = \nabla_{min} - h_{淹} - D_0 - P \quad (m) \tag{10.3}$$

式中：∇_{min} 为出水池最低水位，m。

出水池的净高为

$$\nabla_{池高} = \nabla_{池顶} - \nabla_{池底} \quad (m) \tag{10.4}$$

5. 出水池宽度 B

$$B = (n-1)\delta + n(D_0 + 2a) \quad (m) \tag{10.5}$$

式中：n 为出水管数目；δ 为隔墩厚度，m；D_0 为出水管出口直径，m；a 为出水管边缘至池壁或隔墩的距离，一般取 $a = (0.5 \sim 1.0)D_0$。

6. 出水池长度

（1）水面漩滚法。水平式淹没出流不可避免形成了出水池面层的漩滚（图 10.6），若出水池长度不够，将导致此漩滚延伸至出水干渠，很可能造成渠道的冲刷。漩滚长度与管口淹没深度之间为抛物线的关系，即

$$L_{出} = \alpha h_{淹 max}^{0.5} \quad (m) \tag{10.6}$$

$$\alpha = 7 - \left(\frac{h_p}{D_0} - 0.5\right)\frac{2.4}{1 + \dfrac{0.5}{m^2}} \tag{10.7}$$

式中：$h_{渠 max}$ 为渠道中最大水深，m；α 为试验系数；m 为台坎坡度，$m = h_p/L_p$。

对于垂直台坎，$m = \infty$，此时台坎的影响最大，试验系数为

$$\alpha = 7 - 2.4\left(\frac{h_p}{D_0} - 0.5\right) \tag{10.8}$$

对于水平台坎，$m = 0$，此时台坎无影响，试验系数 $\alpha = 7$。

（2）淹没射流法。图 10.7 为出水池淹没射流示意图。淹没射流法假定出水管管口出流符合半无限空间射流规律，认为水流在池中沿射流方向逐渐扩散，扩散过程中断面平均流速逐渐减小，当断面平均流速等于渠首平均流速时，其扩散长度即为出水池长度。池长计算式为

$$L_{出} = 2.9 D_0\left(\frac{v_0}{v_渠} - 1\right) \quad (m) \tag{10.9}$$

式中：v_0为出水管管口平均流速，m/s；$v_渠$为干渠进口平均流速，m/s。

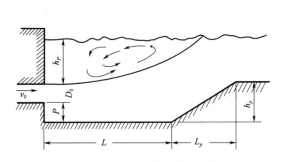

图 10.6 出水池的水面漩滚示意图

Fig. 10.6 Water surface swirl of outlet sump

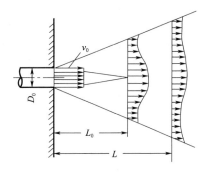

图 10.7 出水池淹没射流示意图

Fig. 10.7 Submerged jet method of outlet sump

实际工程中，通常采用以下修正公式

$$L_出=3.58D_0\left[\left(\frac{v_0}{v_渠}\right)^2-1\right]^{0.41} \quad (\text{m}) \tag{10.10}$$

7. 干渠护砌长度

刚进入干渠的水流紊乱，土渠易被冲刷，故需护砌加固。护砌长度可按下式计算

$$L_护=(4\sim5)h_{渠\max} \tag{10.11}$$

8. 出水池与干渠的渐变段

出水池通常比输水干渠渠底宽，因此，需在两者之间设置一收缩衔接段以实现平顺的过渡。收缩角 α 宜取 $30°\sim40°$，一般不宜大于 $40°$。渐变段长度可按下式计算

$$L_g=\frac{B-b}{2\tan\dfrac{\alpha}{2}} \tag{10.12}$$

9. 侧向出水池尺寸的确定

（1）池宽。对单管侧向出水池，池宽可采用

$$B=(4\sim5)D_0 \quad (\text{m}) \tag{10.13}$$

对多管侧向出流，池宽应随汇入流量的增加而相应加大（图 10.8），不同断面的宽度可按以下方法计算：1-1 断面：$B_1=(4\sim5)D_0$；2-2 断面：$B_2=B_1+D_0$；3-3 断面：$B_3=B_2+D_0$。

（2）池长。图 10.9 所示为单管侧向出水池内的流速分布示意图。当 $L'=5D_0$ 时，流速分布已趋于均匀。所以，单管侧向出水池的池长为

$$L_出=L_2+D_0+L'=L_2+6D_0 \quad (\text{m}) \tag{10.14}$$

图 10.8 所示为多管侧向出水池，池长为

$$L_出=L_2+L_1+L'=L_2+(n+5)D_0+(n-1)S \quad (\text{m}) \tag{10.15}$$

式中：n 为管道数目；S 为管道之间的净距，m。

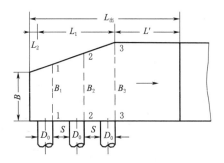

图 10.8　多管侧向出流出水池的尺寸

Fig. 10.8　Dimensions of multi-pipe
side outflow sump

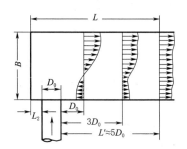

图 10.9　侧向出水池内的流速分布

Fig. 10.9　Flow velocity distribution
in the side outlet sump

10.2　出 水 压 力 水 箱

出水压力水箱也是出水管与干渠之间的连接建筑物，其作用与出水池相同。

出水池是开敞的，压力水箱则是封闭的。当外河水位较高时，出水压力水箱需承受较大的内水压力，因此采用了钢筋混凝土箱形结构。

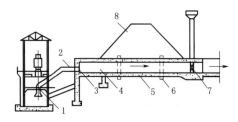

图 10.10　压力水箱的应用

Fig. 10.10　Application of pressure tank

1—水泵 pump；2—出水管 outdoor pipe；3—拍门 flap valve；4—压力水箱 pressure tank；5—压力涵洞 pressure culvert；6—伸缩缝 expansion joint；7—防洪闸 floodgate；8—防洪堤 flood protection dike

出水压力水箱适用于作为堤后式排涝泵站的出水结构。这类泵站外河水位变幅较大，为保证在外河最高水位时也能发挥泵站的排涝作用，若采用开敞式出水池，则势必将出水池修得很高，引起工程量的增加，同时也增加防洪压力；封闭的压力水箱由于尺寸小、工程量省，在这种场合采用相对就比较经济合理。图 10.10 所示为采用压力水箱的泵站出水建筑物示意图。

压力水箱可分为正向出水和侧向出水两种类型，如图 10.11 和图 10.12 所示。与出水池一样，设置隔墩可以有效改善压力水箱内的流态和压力水箱的结构受力条件。

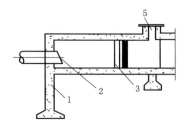
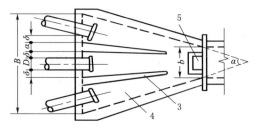

图 10.11　正向出水压力水箱

Fig. 10.11　Forward outlet pressure tank

1—支架 bracket；2—出水口 outlet of pipe；3—隔墩 division pier；4—压力水箱 pressure tank；5—人孔 manhole

出水压力水箱可与泵房分开一段距离，也可紧靠，但无论何种形式，为防止不均匀沉陷导致出水管开裂或压力水箱箱体产生裂缝，为保证泵站的安全运行，压力水箱应建在坚实地基上。如建在填方上，应设置建于原状土上的单独支承。压力水箱与泵房分开浇筑时，出水管道上宜设置柔性接头，以适应站身与压力水箱之间的不均匀沉陷。压力水箱与泵房紧靠时，压力水箱与泵房间应做好止水。

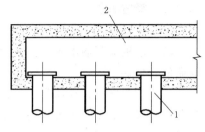

图 10.12 侧向出水压力水箱

Fig. 10.12 Side outlet pressure tank

1—出水管 outdoor pipe；2—压力水箱 pressure tank

压力水箱的箱体一般为钢筋混凝土结构，壁厚30～40cm，隔墩20cm左右。压力水箱的尺寸还应满足检修闸门安装和检修的要求。压力水箱的高度应适宜工作人员进入其内部检修。压力水箱顶部设人孔，其盖板由钢板制成，盖板与人孔之间垫入止水橡皮，并用螺栓紧固，确保盖板的强度和密封性。

10.3 压 力 管 道

出水管道是指水泵出口到出水池之间的输水管道，又称压力管道。中、高扬程泵站往往需要很长的出水管道，管道长度可达数百米甚至更长，其投资在泵站总投资中所占比重很大。因此，正确设计压力管道，对降低工程投资、提高泵站运行效率及安全性都十分重要。

10.3.1 设计要求

（1）管道系统满足稳定性要求。

（2）有足够的管道强度、管道接头强度及密封性，附属设施（通气孔、伸缩节、软接头等）工作可靠。

（3）水力损失小，运行费用低。

（4）经济合理，投资节省。

（5）施工、管理、维修方便。

10.3.2 管线选择

选择管道线路时需考虑下列因素：

（1）管道应铺设在坚实的地基上，地质条件良好，避开松软地基、滑坍地带等地质不良地段和山洪威胁地段；不能避开时，应采取安全可靠的工程措施。

（2）管道布置尽量短，尽可能减少转弯。

（3）控制纵向铺设角度，尽量垂直于等高线布置，注意管道的稳定，防止坍坡、水管下滑等。

（4）管道尽量布置在压坡线以下（压坡线指发生水锤时，管道内水压降低过程线），避免水倒流时管内出现水柱断裂现象，以致引起管道丧失稳定而破坏。

（5）在地形较复杂情况下，可考虑变管坡布置，以减少工程开挖量和避开填方区。压力管道的铺设角一般不应超过土壤的内摩擦角，一般采用1：2.5～1：3的管坡为宜。

10.3.3　管道布置

管道布置形式一般可分为单机单管和多机并联。

（1）单机单管输水的优点是管道结构简单、附件少、运行可靠，一般适用于扬程低、管道短的泵站。

（2）在多机组、高扬程、长管道的泵站中，常采用多机并联方式。这种布置形式可节省管材、减小管床和出水池的宽度，从而减少工程量。但因管道并联，管道附件增多，局部损失加大，年耗电量增加。

在泵站设计中，是否采用多机并联输水，几根管道并联合适，必须进行技术经济比较。

10.3.4　铺设方式

管道的铺设方式有明式和暗式两种。

1. 明式铺设

所谓明式铺设，就是将管道置于露天，如图 10.13 所示。明式铺设的优点是便于管道的安装和检修。缺点是管道因热胀冷缩，来回滑动频繁，缩短管道的使用寿命。对于钢筋混凝土管，因管壁直接受太阳辐射等气温影响，在夏季，管壁内外温差较大，温度应力有时很大，如忽略有可能产生裂缝。在严寒地区冬季运行时，可根据需要对管道采取防冻保温措施；若冬季不运行，要将管道中的水放空或采取相应的防护措施，以防管道冻坏。

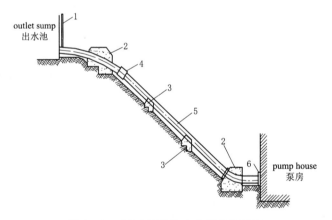

图 10.13　明式铺设的出水管道示意图

Fig. 10.13　Open laying pipeline

1—通气管 ventilator；2—镇墩 anchor block；3—支墩 buttress；4—伸缩节 expansion joint；

5—钢管 steel pipe；6—穿墙软接头 soft joint through wall

金属管道多采用明式铺设，为防止锈蚀，其外壁应刷漆保护。

2. 暗式铺设

暗式铺设是将管道埋于地下，其优点是管道受温度变化影响小，在严寒地区冬季仍可输水，但安装和检修费用较高。暗式铺设管底最小埋深应在最大冻土深以下；采用连续垫座，埋入地下的金属管道应做防锈处理和防侵蚀措施；管道穿过河床时其埋置深度应考虑河床冲刷的影响；对于南方的非冻土区，管顶埋设深度主要取决于外部荷载。

在选择管道铺设方式时，应根据管材、供水情况等因素综合考虑确定。

10.3.5 管材、管径及附件

压力管道适用的管材较多，包括铸铁管、钢管、钢筋混凝土管及预应力钢筋混凝土管等。管材的选择对管道的运输、安装、维护、管理、投资等都有很大影响，故需综合考虑各方面的因素。

1. 铸铁管

铸铁管可分为低压管（4.5×10^5 Pa）、中压管（7.5×10^5 Pa）和高压管（10×10^5 Pa）三个压力等级，具有价格低廉、安装方便、不易腐蚀的优点和性脆、壁厚、笨重的缺点，适用于管径小于 600mm 的出水管道。

2. 钢管

钢管具有强度高、管壁薄、重量轻、接头简单、运输方便等优点，缺点是易腐蚀、使用期限短，适于高扬程泵站和管径大于 800mm 的出水管道。

3. 钢筋混凝土管

钢筋混凝土管价格低廉、使用期长、运行管理费用低、输水性能好，但运输不便，承插接头处有时因预制误差较大或安装时防漏处理不严而引起漏水。钢筋混凝土管适用于管径 300～1500mm 的低压管道。

4. 预应力钢筋混凝土管

预应力钢筋混凝土管的突出优点是节省钢材，具有较高的弹性和抗渗抗裂性，并能承受较大的内水压力，应用较为广泛。

5. 其他材质管

聚氯乙烯管具有重量轻、便于运输和安装、水力损失小、不锈蚀等优点。选用聚氯乙烯管需要暗式铺设。

此外，也有不少泵站采用玻璃钢管。

管径大小直接影响泵站投资和泵站效率，管径增大，管道阻力减小，可降低耗电量，但管道一次性投资相应增加；相反，管径减小，虽可降低管道投资，但管道的阻力增大，年耗电量增加。因此，如何确定经济合理的管径，应有一个最优方案，即经济管径。

在初步设计阶段，也可根据经济流速确定经济管径。泵站出水管经济流速，一般净扬程 50m 以下取 1.5～2.0m/s，净扬程 50～100m 时可取 2.0～2.5m/s。

出水管道的安装和安全运行，要附设一些必需的附件及设备，如闸阀、伸缩节、拍门、通气孔等。管道附件因管道长短、扬程高低、铺设方式的不同而变，应该安装哪些附件，由泵站的具体情况而定。

10.3.6 出水管道支承结构

为固定出水管道并维持其稳定，消除正常运行及事故停机时产生的振动和位移，泵站出水管道必须设置稳固的支承结构，其形式可分为支墩和镇墩两种。

支墩主要用于长管段，其作用是承受管道及水的重力、减少振动。支墩的断面尺寸按构造设计即可（图 10.14 和图 10.15）。除伸缩节附近处，其他各支墩宜采用等间距布置。

钢管的支墩间距一般可取 5～10m；预应力钢筋混凝土管道应采用连续管座或每节设两个支墩。预制钢筋混凝土管长度大多为每节 5m，且采用承接式接头，每节宜设两个支墩，分别设在每节管的 1/4 和 3/4 处。

镇墩主要用于管道转弯处，也用于斜坡上的长管段，一般间隔 80～100m 设一镇墩，其主要作用是承受管道转弯处的各种作用力，抵消坡道上由重力引起的下滑力。管道转弯处必须设置镇墩，在明管直线段上镇墩的间距不宜超过 100m。两镇墩之间应设伸缩节，伸缩节应布置在上端。镇墩的断面尺寸可通过具体受力分析和结构计算确定。

镇墩有两种形式：封闭式和开敞式。封闭式将管道设于镇墩之内，如图 10.16（a）所示；开敞式将管道置于镇墩表面，如图 10.16（b）所示。封闭式镇墩与管道的固定较为牢固，开敞式镇墩则便于检查与维修。大中型泵站出水管道转弯处受力情况较为复杂，常用封闭式镇墩。

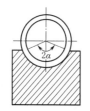

图 10.14 混凝土支墩

Fig. 10.14 Concrete buttress

图 10.15 浆砌块石支墩

Fig. 10.15 Masonry block stone buttress

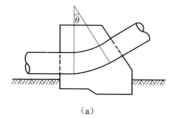 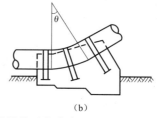

（a） （b）

图 10.16 管道镇墩的两种形式

Fig. 10.16 Two forms of pipe anchor block

（a）封闭式；（b）开敞式

（a）closed；（b）open

10.4 出 水 流 道

出水流道是连接水泵导叶出口与出水池的衔接通道，多应用于扬程较低的大中型泵站。出水流道一般采用钢筋混凝土现浇，在结构上与泵房连成整体，具有长度较短、断面形状变化较大、水力损失较小的特点。出水流道有多种形式，如虹吸式出水流道、直管式出水流道、斜式出水流道等。

10.4.1 出水流道的作用与设计要求

出水流道的作用是使水流在从水泵导叶出口流入出水池的过程中更好地转向和扩散，

在不发生脱流或漩涡的条件下最大限度地回收动能。出水流道内的流态及动能回收情况决定了出水流道的水力损失。对于低扬程泵站，出水流道水力损失在水泵总扬程中所占的比例较大，对泵装置的能量性能有较为明显的影响。由此可见，出水流道也是水泵装置的一个重要组成部分。

《泵站设计规范》对出水流道的设计提出如下要求：

(1) 与水泵导叶出口相连的出水形式应根据水泵的机构和泵站的要求确定。

(2) 流道型线变化应比较均匀，当量扩散角宜取 8°～12°。

(3) 流道出口流速不宜大于 1.5m/s（出口装有拍门时不宜大于 2.0m/s）。

(4) 应有合适的断流方式。

(5) 平直管出口宜设置检修门槽。

(6) 应施工方便。

在上述要求中，第（2）项要求是为了水流扩散均匀和避免产生脱流或漩涡；第（3）项要求是为了尽量减少流道出口的动能损失；第（4）项要求说明对断流方式应予以足够重视。在此基础上，可适当兼顾减少土建投资及施工方便等其他方面的要求。

10.4.2 典型出水流道

1. 虹吸式出水流道

虹吸式出水流道（图 10.17）适用于出水池水位变化不大的立式或斜式低扬程泵站，其主要优点是运行方便可靠，水泵停机时可通过真空破坏阀破坏虹吸、切断水流。此外，虹吸式出水流道还便于穿越堤防，不影响防洪堤的安全。我国目前已有上百座大型泵站采用了虹吸式出水流道。虹吸式出水流道的缺点是：工程量较大、施工较为困难、设计不当易引起机组振动等。

虹吸式出水流道的工作原理如图 10.18 所示。水泵启动前，虹吸管内高出水面以上的部分充满空气，水泵启动后，出水流道内的水位迅速上升，此时，流道内的空气被压缩，当流道内的压力 P_c 高于大气压力 P_a 一定值时［图 10.18 (a)］，设于驼峰顶部的真空破坏阀被打开，流道内的部分空气将通过真空破坏阀被排出流道［图 10.18 (b)］。当水泵提升的水位超过驼峰底部时，水流就会越过驼峰，像溢流堰那样沿管壁下泄，同时还

图 10.17　虹吸式出水流道
Fig. 10.17　Siphon outlet passage

会挟带管道内的剩余空气从流道出口流出。待流道内的空气被全部排出、水流充满全流道后，水泵的启动过程便告结束，进入正常运行状态［图 10.18 (c)］。

由于虹吸式出水流道驼峰顶部的高程高于最高出水位，在形成满管流以后，流道驼峰附近必为负压。可见，这种出水流道利用了虹吸原理。虹吸作用形成的过程实质就是在水泵启动过程中水流充满管段、空气排出管外、在驼峰处形成一定真空的过程。当水泵正常停机或事故停泵时，可及时打开真空破坏阀，利用驼峰处的负压，使空气进入流道顶部、破坏流道内的真空［图 10.18 (d)］，从而达到截断水流的目的。

合理设计流道，既要减少流道水力损失，又要保证启动过程中具有较强的挟气能力，以尽可能地缩短虹吸形成的时间。

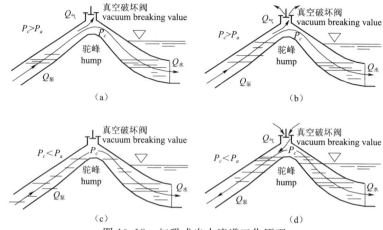

图 10.18 虹吸式出水流道工作原理

Fig. 10.18 Working principle of the siphon outlet passage

(a) 启动；(b) 排气；(c) 运行；(d) 停机

(a) start；(b) air-out；(c) operation；(d) shutdown

图 10.19 直管式出水流道

Fig. 10.19 Straight pipe outlet passage

2. 直管式出水流道

直管式出水流道（图 10.19）与水泵出口弯管相接，流道断面形状由圆变方，在平面方向和立面方向均逐渐扩大，流道内的平均流速逐渐减小，流道内任一断面都具有一定正压力。直管式流道断面形状简单、施工方便，采用拍门和快速闸门作为断流措施，在大中型泵站中已有较多应用。

为了避免不必要的能量损失，出水流道的出口应淹没在出水池最低运行水位以下，出水流道出口上缘的最小淹没深度宜取 0.3～0.5m。根据水泵出水弯管出口断面中心高程和出水池最低水位的相对尺寸，直管式出水流道有上升式、平管式和下降式三种布置形式，如图 10.20 所示。

直管式出水流道都应设有通气孔，其目的是机组在启动阶段可以由通气孔排气，在停机阶段可以由通气孔补气，以减弱流道内的压力脉动。

通气孔的布置决定于流道的布置形式，对于下降式流道或弯曲的低驼峰流道，通气孔应布置在流道最高位置，对于上升式则可布置在流道出口附近。

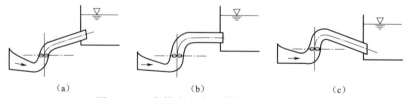

图 10.20 直管式出水流道的三种布置形式

Fig. 10.20 Three layouts of straight pipe outlet passage

(a) 上升式；(b) 平管式；(c) 下降式

(a) ascending；(b) flat；(c) descending

3. 斜式出水流道

斜式出水流道与斜式轴伸泵装置配套使用，其
进口与水泵导叶出口直接相接，水泵轴线与水平方
向的夹角 α 常取为 45°、30°或 15°，如图 10.21 所
示。斜式出水流道可分为弯曲段和直线段两个部
分。弯曲段断面形状由圆变方，其进口断面为圆
形，出口断面为矩形，在平面方向和立面方向均逐
渐扩大，流道内的平均流速逐渐减小。由于斜式轴

图 10.21　斜式出水流道
Fig. 10.21　Oblique outlet passage

伸泵装置的驱动装置位于出水流道的上方，为了给电机及其散热风道留下必要的空间，弯
曲段向下弯曲得很厉害。泵轴与水平方向的夹角 α 越小，弯曲得越厉害。直线段断面形状
均为矩形，在平面方向和立面方向的尺寸均以线性变化的方式逐步扩大，流道内的平均流
速逐渐减小。

与直管式出水流道一样，斜式出水流道内任一断面都具有一定正压，在断流方式及通
气孔方面的要求也与直管式出水流道相同。

10.4.3　出水流道的水力设计

1. 一维水力设计方法

出水流道一维水力设计方法与进水流道的类似，以一维流动理论为基础，其要点可概
括为：

(1) 假定断面平均流速等于设计流量除以断面面积。

(2) 要求沿流道断面中心线的各断面平均流速均匀变化。

根据经验数据初步拟定出水流道的主要尺寸后，就可以根据流速均匀变化的要求绘出
流道的纵剖面轮廓图，然后再按流速递减法绘制平面轮廓图，具体步骤为：

(1) 先初定一个平面轮廓图，在剖面轮廓图和平面轮廓图中选取若干个断面，由剖面
图得到各断面的高度，由平面图得到断面的宽度，从而求出各断面的面积。

(2) 由设计流量求各断面的平均流速。

(3) 做出平均流速和流道长度、断面面积和流道长度的关系曲线。若上述两条曲线光
滑，则符合要求；否则应重新调整剖面或平面图尺寸，重复步骤 (1) ～ (3)，直至满足
要求。

2. 三维优化水力设计方法

出水流道三维优化水力设计方法的基本思路是：在给定控制尺寸的条件下，给定不同
的出水流道边界，完成相应的流场 CFD 计算，考察不同边界时流道内的三维流态，以流
道内不发生脱流和漩涡、流道水力损失尽可能小为目标，逐一优化流道几何参数、调整流
道型线，逐步实现流道的最佳水力性能。

10.4.4　断流方式

1. 拍门

拍门是一种单向阀门，多与直管式、斜式等形式的出水流道配套使用，是最常见的一
种断流方式。水泵开机后，拍门在水流的冲击下自动打开；水泵停机后，靠拍门的自重及
水流反向流动的作用力自动关闭。拍门的门顶用铰链与门座相连，门与门座之间用橡皮止

水，其主要特点是结构简单、应用方便、造价低廉，在泵站中得到了广泛的应用。

普通拍门没有任何附加控制设备，靠水流的冲击打开，靠自重或反向水流的作用力关闭，多用于中小型泵站〔图 10.22 （a）〕。由于拍门是在水流的冲击下开启的，所以在正常运行时经常需要消耗一定的能量。为了减少拍门阻力，有些大型泵站采用了机械平衡式拍门、浮箱式拍门等〔图 10.22 （b）〕。另外，由于拍门是在自重或反向水流的作用下关闭的，其在最后关闭阶段速度较大，从而会产生较大的撞击力。为了避免较大撞击力对拍门及门座有可能造成的破坏，有些泵站采用了带有液压缓冲装置的拍门〔图 10.22 （c）〕。

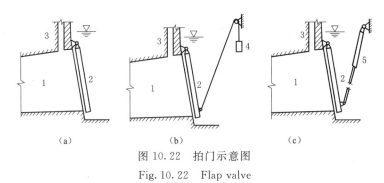

图 10.22　拍门示意图

Fig. 10.22　Flap valve

（a）普通拍门；（b）带平衡锤的拍门；（c）带液压缓冲装置的拍门

（a）normal flap valve；（b）flap valve with counterweight hammer；（c）flap valve with hydraulic buffer

1—出水流道 outlet passage；2—拍门 hap valve；3—门座 hap valve seat；

4—平衡锤 balance hammer；5—液压缓冲装置 hydraulic buffer

2. 快速闸门

快速闸门通常配用液压启闭机，能快速启闭，是大型泵站的又一种断流方式，适用于直管式、斜式等形式的出水流道。这种断流方式的显著优点是启闭迅速、撞击力小。

快速闸门的“快速”主要是在水泵停机时，要求闸门迅速下落、截断水流。在水泵机组启动时，若闸门开启太慢，对轴流泵机组而言，就会增加水泵的启动扬程，从而导致电机过载及机组振动；如果闸门开启太快，则可能使水泵排出的水和从闸门外流进的水在流道内相撞，从而导致流道内排气困难、引起较大压力脉动。为保证机组在启动过程中避免产生由于闸门开启过快或过慢而引发的问题，可采取一定的安全措施，如设胸墙或在快速闸门的门页上开小拍门等，如图 10.23 所示。采取安全措施后，对于快速闸门的开启时间和速度就可不必严格要求。

3. 真空破坏阀

真空破坏阀与虹吸式出水流道配套，用以破坏虹吸流道顶部的真空，隔断水流。在水泵机组停机以后，只要把设置在驼峰顶部的阀门打开，使空气进入流道，就可以破坏真空、截断水流。这种断流方式的显著优点是操作简便、断流可靠、检修方便。为了保证机组正常和安全运行，真空破坏阀应满足密封性能好、动作迅速可靠、放气灵敏度高等要求。

最常用的真空破坏阀是气动平板阀，由阀座、阀盖、气缸、活塞、活塞杆、弹簧等部分组成，如图 10.24 所示。如真空破坏阀因故不能打开，还可以打开手动阀，将压缩空气送入真空破坏阀的气缸，使真空破坏阀强制动作打开。如因特殊原因真空破坏阀无法打开，可用大锤击破真空破坏阀旁的有机玻璃板，使空气进入虹吸管内，保证在水泵停机后能可靠地破坏虹吸管的真空、截断水流。

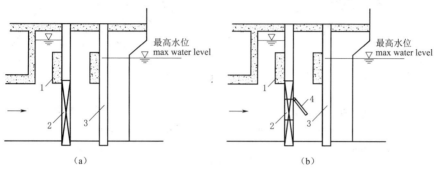

图 10.23　快速闸门示意图

Fig. 10.23　Schematic diagram of the rapid-drop gate

（a）采用胸墙溢流的快速闸门；（b）带小拍门的快速闸门

（a）A fast gate with a chest wall overflow；（b）A fast gate with a small flap valve

1—胸墙 chest wall；2—快速闸门 rapid-drop gate；3—检修门 overhaul door；4—小拍门 small flap valve

　　（1）水力冲动式。水力冲动式真空破坏阀如图 10.25 所示。水泵正常运行时，水流沿箭头方向流动，在水流冲击下挡板杠杆绕支点转动，从而使阀板压紧进气口，使管内保持一定真空度。当水泵停止运行后，水流倒流，挡板受倒流作用使阀杆绕支点反向转动，打开进气口，空气进入管道，破坏真空，起断流作用。

　　（2）通气管式。通气管式真空破坏阀破坏真空的方法是把进气口从驼峰顶处引到上游低于驼峰顶高程的一定位置，如图 10.26 所示。这样，在水泵正常运行时，因上游压力高，水位上升，淹没进气口，起到密封作用，以利真空形成。在水泵停机，水流产生倒流后，原上游的水压力下降，水箱内的水位随之下降，当水位下降到进气口以下时，空气即进入管道从而破坏真空、切断水

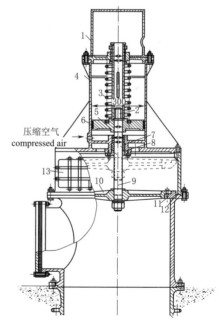

图 10.24　真空破坏阀结构图

Fig. 10.24　Vacuum breaking valve structure

1—罩壳 casing；2—活塞杆 piston rod；3—弹簧 spring；
4—气缸 cylinder；5—活塞 piston；6—活塞环 piston ring；
7—填料 packing；8—填料压盖 packing gland；9—阀杆 valve
stem；10—阀门 valve；11—阀门外环 valve outer ring；
12—阀门座 valve seat；13—进气滤网 inlet filter

213

流。这种真空破坏阀结构较简单，但受出口水位变化限制，出口水位变幅不能太大。

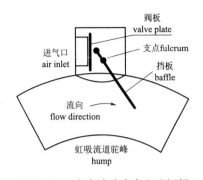

图 10.25　水力冲动式真空破坏阀

Fig. 10.25　Hydraulic impulse vacuum breaking valve

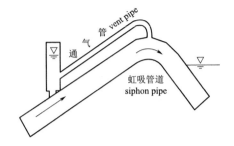

图 10.26　通气管式真空破坏阀

Fig. 10.26　Vented pipe vacuum breaking valve

Chapter 10　Outlet Structures

Outlet structures of pumping station mainly include outlet sump (or pressure tank), outlet passage, outdoor pipeline, outlet channel and cut-off facilities. It must meet the following requirements:

(1) Ensure sufficient stability and safety.

(2) Good flow pattern, maximum recovery of flow kinetic energy at pump outlet.

(3) Joints and seals of all parts are in good condition.

(4) Venting and damaging vacuum facilities during water discharge and shut-off.

(5) Convenient construction and operation management.

(6) Save civil investment.

10. 1　Outlet sump

10. 1. 1　Function and design requirements of outlet sump

The outlet sump is a connecting building connecting the outdoor pipeline and the outlet channel, as shown in Fig. 10. 1. Its main functions are:

(1) Collect the inflow of the outdoor pipeline and sometimes act as a shunt to several main channels connected to the outlet sump.

(2) Prevent erosion and stabilize flow, diffuse outflow, and draw water smoothly into the main channel to avoid channel erosion.

(3) It is convenient to set up facilities to prevent backflow when the pump is stopped and to set up facilities for overhaul and cut-off.

The position of the outlet sump is generally higher than that of the pump house. Once the collapse accident occurs, the safety of the whole pumping station will be endangered. Therefore, the structure of the outlet sump must be reliable and stable, and it should be built on the excavation. The seepage prevention calculation, foundation pressure check and foundation settlement calculation of the outlet sump must meet the requirements of the specification. The height of the outlet sump must ensure that the maximum discharge does not overflow and ensure stability and reliability. The flow velocity is less than 2m/s. In order to prevent the channel from being scoured, the main channel entrance should have a certain length of masonry.

10. 1. 2　Type of outlet sump

The types of outlet sumps are generally classified into three types according to the flow direction: direct outlet sump, side outlet sump and multi-directional diversion outlet sump.

(1) The flow direction at the nozzle of the direct outlet sump is the same as that of the main channel [Fig. 10. 2 (a)] . The flow direction is straight, the hydraulic performance is good, and it is widely used.

(2) The flow direction at the nozzle of the side outlet sump is orthogonal to that of the main channel [Fig. 10. 2 (b)] . Since the flow turns 90 degree, the flow crosses and collides with the wall of the sump, and the hydraulic performance is poor, it is usually only adopted when the direct outlet sump can not be arranged.

(3) The multi-directional diversional outlet sump is used for simultaneous water transfer to several main channels. The flow pattern in the sump is between the direct and side outlet sump [Fig. 10. 2 (c)] . The hydraulic performance can be improved by taking appropriate diversion measures.

According to the relationship between outlet sump and pump house, it can be classified into two types: split-type and joint-type. The former separates outlet sump from pump house and is mostly used for high-head pumping station, while the latter integrates outlet sump and pump house and is mostly used for low-head pumping station.

According to the different ways of cut-off, it can be classified into different types of outlet sumps, such as slap-gate type, gate type, siphon type, overflow weir type and free-flow type.

According to whether there is free water surface or not, it can be classified into open outlet sump and pressure tank. The former is commonly referred to as outlet sump, with free water surface in the sump, and the latter is a closed pressurized pipe.

10. 1. 3　Flow pattern of outlet sump

Fig. 10. 3 show the flow pattern of the direct outlet sump. After entering the outlet sump from the outdoor pipe, water diffuses simultaneously in the elevation direction and the plane direction, presenting a limited space three-dimensional diffusion submerged jet state. In the elevation direction, there is a large swirl zone in the upper part of the main stream and a small swirling area in the lower part; in the plane direction, a backflow area is formed on both sides of the outdoor pipe.

Due to the existence of swirl zone, the flow in the sump is disordered, and improper treatment will lead to increased hydraulic loss and scour of outlet sump and channel. The shape and size of the swirl zone (including the elevation diffusion angle alpha, the angle of plane diffusion α and the length of swirl L in the Fig. 10. 3) are closely related to the flow of the outdoor pipe, the submerged depth of the pipe mouth and the geometric boundary of the outlet sump.

Fig. 10. 4 shows the flow pattern of the side outlet sump. The flow in the outlet sump is blocked by the direct wall to form a reverse backflow, and the outflow is not smooth, resulting in high water level and increased hydraulic loss. The closer the wall is to the orifice, the greater the influence of outflow flow pattern.

10. 1. 4   Determination of outlet sump dimensions

Fig. 10. 5 shows a submerged flow pattern of direct outlet sump, in which the flow from the outdoor pipe to the outlet sump horizontally. The definition of geometric parameters of the outlet sump is given in the Fig. 10. 5.

1. Outlet diameter of outdoor pipe

In order to reduce the outlet velocity, avoid hydraulic jump in the outlet sump and reduce the outlet loss, the outlet diameter of the outdoor pipe should be larger; on the other hand, the outlet diameter of the outdoor pipe should not be too large, so as to avoid excessive increase of the matching flap valve and relevant size. Generally, the average velocity of outdoor pipe orifice is selected in the range of 1. 5 – 2. 5m/s.

2. Submergence depth

A certain submergence depth should be reserved at the outdoor pipe orifice, so as not to make the water flow out of the water surface, increase the hydraulic loss and water surface whirling. The value of the submergence depth is related to the velocity of the pipe orifice. It is usually taken as

$$h_{淹} = (2\sim3)\ \frac{v_0^2}{2g}\ (m) \tag{10.1}$$

3. Distance from bottom of sump to lower edge of outdoor pipe

In order to facilitate the installation of outdoor pipe and flap valve, and avoid blockage of pipe orifice by sediment or debris, there should be a certain space between outdoor pipe orifice and outlet sump bottom, generally $P > 0.3$m.

4. Elevation of top and bottom of outlet sump wall

The height of the outlet sump shall be such that no overflow occurs at the maximum water level. The top of wall elevation of the outlet sump is calculated as follows

$$\nabla_{池顶} = \nabla_{max} + h_{超高}\ (m) \tag{10.2}$$

Where, $h_{超高}$ is safety swper elevation, which is related to the discharge of pumping station, and can be selected by referring to Table 10. 1; ∇_{max} is the maximum water level of the outlet sump.

The floor elevation of outlet sump is

$$\nabla_{池底} = \nabla_{min} - h_{淹} - D_0 - P\ (m) \tag{10.3}$$

Where, ∇_{min} is the minimum water level of the outlet sump.

The net height of outlet sump is

$$\nabla_{池高} = \nabla_{池顶} - \nabla_{池底}\ (m) \tag{10.4}$$

5. Outlet sump width

$$B = (n-1)\ \delta + n\ (D_0 + 2a)\ (m) \tag{10.5}$$

Where, n is the number of outdoor pipes; δ is the thickness of division pier, m; D_0 is the outlet diameter of outdoor pipes, m; a is the distance from the edge of outdoor pipes to the sump wall or division pier, and generally $a=$ (0.5 – 1.0) D_0 is taken.

6. Length of outlet sump

(1) Surface vortex method. The horizontal submerged outflow will inevitably form vortices on the surface of the outlet basin (Fig. 10.6) . If the length of the outlet sump is not enough, it will extend to the main outlet channel, which is likely to cause channel erosion. The relationship between vortex length and submerged depth is parabola

$$L_{出}=ah_{淹max}^{0.5} \quad (m) \tag{10.6}$$

$$\alpha=7-\left(\frac{h_p}{D_0}-0.5\right)\frac{2.4}{1+\frac{0.5}{m^2}} \tag{10.7}$$

Where, $h_{渠max}$ is the maximum water depth in the channel, m; α is the test coefficient; m is the slope of the platform sill, $m=h_p/L_p$.

For vertical sill, $m=\infty$, the influence of the sill is the largest, and the test coefficient is

$$\alpha=7-2.4\left(\frac{h_p}{D_0}-0.5\right) \tag{10.8}$$

For horizontal sill, $m=0$, the sill has no effect, and the test coefficient $\alpha=7$.

(2) Submerged jet method. Fig. 10.7 is a schematic diagram of submerged jet in the outlet sump. The submerged jet method assumes that the outflow at the outlet of the outdoor pipe conforms to the semi-infinite space jet law. It considers that the flow diffuses gradually along the jet direction in the sump, and the average cross-section velocity decreases gradually during the diffusion process. When the average cross-section velocity is equal to the average flow velocity at the head of the channel, its diffusion length is the length of the outlet sump. The sump length calculation formula is

$$L_{出}=2.9D_0\left(\frac{v_0}{v_渠}-1\right) \quad (m) \tag{10.9}$$

Where, v_0 is the average velocity at the outlet of the outdoor pipe, m/s; $v_渠$ is the average velocity at the inlet of the outlet channel, m/s.

The following correction formulas are usually used in practical engineering projects

$$L_{出}=3.58D_0\left[\left(\frac{v_0}{v_渠}\right)^2-1\right]^{0.41} \quad (m) \tag{10.10}$$

7. Length of masonry of main channel

The flow entering the main channel is disordered and the soil channel is easily washed out, so it needs to be reinforced by masonry. The length of masonry can be calculated as follows

$$L_{护}= (4\sim5)\ h_{渠max} \tag{10.11}$$

8. Transition section of outlet sump and trunk channel

The outlet sump is usually wider than the bottom of the main channel for water conveyance. Therefore, a contraction connection section is required between the two to achieve a smooth transition. The shrinkage angle α should be from $30°$ to $40°$, but should not be greater than $40°$ in general. The length of the transition section can be derived from the following formula

$$L_g = \frac{B-b}{2\tan\dfrac{\alpha}{2}} \tag{10.12}$$

9. Dimensions of side outlet sump

(1) Sump width. For single pipe side outlet sump, the sump width can be adopted

$$B = (4\sim5)\ D_0\ (m) \tag{10.13}$$

For multi-pipe side outlet sump, the sump width should increase with the increase of discharge (Fig. 10. 8), and the width of different sections can be calculated as follows: section 1 - 1: $B_1 = (4\sim5)\ D_0$; section 2 - 2: $B_2 = B_1 + D_0$; section 3 - 3: $B_3 = B_2 + D_0$.

(2) Sump length. Fig. 10. 9 shows the flow velocity distribution in the single-pipe side outlet sump. When $L' = 5D_0$, the flow velocity distribution has tended to be uniform. Therefore, the sump length of the single-pipe side outlet sump is

$$L_{出} = L_2 + D_0 + L' = L_2 + 6D_0\ (m) \tag{10.14}$$

Fig. 10. 8 shows a multi-pipe side outlet sump with the length of

$$L_{出} = L_2 + L_1 + L' = L_2 + (n+5)\ D_0 + (n-1)\ S\ (m) \tag{10.15}$$

Where, n is the number of pipes; S is the net distance between different pipes, m.

10. 2　Outlet pressure tank

The outlet pressure tank is also the connecting building between the outdoor pipe and the main channel, which plays the same role as the outlet sump.

The outlet sump is open and the outlet pressure tank is closed. When the water level of the outer river is higher, the outlet pressure tank needs to bear greater internal water pressure, so the reinforced concrete box structure is adopted.

The outlet pressure tank is suitable for the outlet structure of the pumping station for drainage of waterlogging behind the dike. In order to ensure that the water level of the pumping station can also play a drainage role when the highest water level is in the outer river, if open outlet sump is used, the outlet sump will inevitably be designed very high, causing an increase in the amount of projection, and also increasing the flood control pressure. The closed pressure tank is relatively economical and reasonable in this case because of its small size and low amount of projection. Fig. 10. 10 shows the schematic diagram of the pumping station outlet structure with pressure tank.

The outlet pressure tank can also be classified into two types: forward outlet and side outlet, as shown in Fig. 10. 11 and Fig. 10. 12. As with the outlet sump, the setting of piers can effectively improve the flow pattern in the outlet pressure tank and the structural stress conditions of the outlet pressure tank.

The outlet pressure tank can be separated from or close to the pump house for a certain distance, but in any form, in order to prevent uneven subsidence leading to cracking of the outdoor pipe or cracking of the outlet pressure tank body, in order to ensure the safe operation of the pumping station, the outlet pressure tank should be built on a solid foundation. If constructed on filling, separate supports shall be provided on undisturbed soil. When the outlet pressure tank is poured separately from the pump house, flexible joints should be installed in the outdoor pipe to adapt to the uneven settlement between the station body and the outlet pressure tank. When the outlet pressure tank is close to the pump house, the outlet pressure tank and the pump room should be well watertight.

The box of pressure tank is generally reinforced concrete structure with wall thickness of 30 – 40cm and division pier of about 20cm. The size of the outlet pressure tank shall also meet the requirements for the installation and maintenance of the overhaul gate. The height of the outlet pressure tank should be suitable for staff to enter its internal maintenance. The top of the outlet pressure tank is provided with a manhole, and the cover plate is made of a steel plate. A water stop rubber is padded between the cover plate and the manhole, and is fastened with bolts to ensure the strength and sealing of the cover plate.

10. 3　Pressure pipeline

The outdoor pipeline refers to the water pipeline between the outlet of the pump and the outlet sump, also known as the pressure pipeline. Medium and high head pumping stations often need a long water outdoor pipeline, and the length of the pipeline can reach hundreds of meters or even longer. Its investment accounts for a large proportion of the total investment of pumping stations. Therefore, the correct design of pressure pipeline is very important to reduce the project investment, improve the operation efficiency and safety of the pumping station.

10. 3. 1　Design requirements

(1) The pipeline system meets the stability requirements.

(2) Adequate pipe strength, pipe joint strength and sealing, and reliable operation of auxiliary facilities (vents, expansion joints, soft joints, etc.).

(3) Low hydraulic loss and operating cost.

(4) Economic and reasonable, investment saving.

(5) Convenient construction, management and maintenance.

10. 3. 2　Selection of pipeline

The following factors should be considered when choosing pipeline route:

(1) Pipelines should be laid on solid foundations with good geological conditions to a-

void unfavorable geological sections such as soft foundations, sliding zones and areas threatened by mountain torrents; when they cannot be avoided, safe and reliable engineering measures should be taken.

(2) Pipeline layout should be as short as possible to minimize turning.

(3) Control the longitudinal laying angle, arrange perpendicular to the contour line as far as possible, pay attention to the stability of the pipeline, prevent the collapse, water pipe sliding, etc.

(4) The pipeline should be arranged below the gradient line as far as possible (gradient line refers to the process line of reducing water pressure in the pipeline when water hammer occurs), so as to avoid water column fracture in the pipeline when the backflow occurs, so as to cause the unstability and damage of the pipeline.

(5) In case of complex topography, the layout of variable pipe slope can be considered to reduce the amount of excavation and avoid filling area. The laying angle of pressure pipeline should not exceed the internal friction angle of soil in general, and the pipe slope of $1:2.5 - 1:3$ should be adopted generally.

10. 3. 3   Pipeline layout

The pipeline layout can be generally classified into single-units single-pipe and multiple units parallel connection.

(1) The advantages of single-units single-pipe pipeline layout are simple pipeline structure, few accessories and reliable operation, which are generally applicable to pumping stations with low head and short pipeline.

(2) In pumping stations with multiple units, high head and long pipeline length, multiple units parallel connection is often adopted. This arrangement can save pipes, reduce the width of pipe bed and outlet sump, and thus reduce the amount of work. However, due to parallel connection of pipelines, pipeline accessories increase, local losses increase, and annual power consumption increases.

In the design of pumping station, it is necessary to make technical and economic comparison whether multi-machine parallel water transfer and several pipelines are suitable for combination.

10. 3. 4   Pipeline laying mode

There are two laying modes of pipeline, i. e. open and closed.

1. Open laying

The so-called open laying is to place the pipeline in the open air, as shown in Fig. 10. 13. The advantages of open laying are easy installation and maintenance of pipelines. The disadvantage is that the pipeline slides back and forth frequently due to thermal expansion and cold contraction, which shortens the service life of the pipeline. For reinforced concrete pipe, because the shell of pipe is directly affected by temperature such as solar radiation, in summer, the temperature difference between the inside and outside of the pipe shell is large, and the temperature stress is sometimes large, if neglected, cracks may occur. When operating in winter in severe cold areas, anti-freezing and heat preservation measures can be taken for pipelines according to needs; if not operating in winter, water in pipelines should be emptied or corresponding protective measures should be taken to prevent pipeline frost damage.

Metal pipes are mostly laid in the open style. In order to prevent rust, their outer shell should be painted and protected. Fig. 10. 14 shows a schematic diagram of the outdoor pipe laid in the open style.

2. Closed laying

Closed laying is to bury the pipeline underground. The advantage is that the pipeline is less affected by temperature changes and can still carry water in winter in severe cold areas, but the installation and maintenance costs are high. The minimum burial depth of closed laying pipe bottom shall be below the maximum frozen soil depth; continuous cushion shall be used; metal pipelines buried underground shall be rust-proof and corrosion-proof measures; the burial depth of pipelines when crossing the riverbed shall consider the impact of riverbed scour; for non-frozen soil areas in the south, the burial depth of pipe top mainly depends on external load.

When choosing pipeline laying mode, it should be determined comprehensively according to factors such as pipe material, water supply and other factors.

10. 3. 5   Pipeline materials, diameters and accessories

There are many applicable pipes for outdoor pipelines, including cast iron pipes, steel pipes, reinforced concrete pipes and prestressed reinforced concrete pipes. The selection of pipe materials has a great impact on the transportation, installation, maintenance, management, investment and so on of the pipeline, so all aspects of factors need to be considered comprehensively.

1. Cast iron pipe

The cast iron pipes can be classified into three pressure classes: low pressure pipes (4.5×10^5 Pa), medium pressure pipes (7.5×10^5 Pa) and high pressure pipes (10×10^5 Pa). They have the advantages of low price, convenient installation, non-corrosion, and the disadvantage of brittleness, thick wall and bulky, and are suitable for outdoor pipes with diameters less than 600mm.

2. Steel pipe

The steel pipes have the advantages of high strength, thin wall, light weight, simple joints and convenient transportation. Their disadvantages are easy corrosion, short service life, and are suitable for high-head pumping stations and outdoor pipes with diameters greater than 800mm.

3. Reinforced concrete pipe

The reinforced concrete pipes are inexpensive, have long service life, low operation and management costs, and have good water transmission performance, but the transportation is inconvenient. Sometimes, the leakage occurs at socket joints due to large prefabrication errors or inadequate leakage prevention treatment during installation. The reinforced concrete pipes are suitable for low pressure pipes with a diameter of 300mm to 1500mm.

4. Prestressed reinforced concrete pipe

The outstanding advantages of prestressed reinforced concrete pipe are steel saving, high elasticity, seepage resistance and crack resistance, and can withstand large internal water pressure, so it is widely used.

5. Pipeline of other materials

The polyvinyl chloride (PVC) pipes have the advantages of light weight, convenient transportation and installation, small hydraulic loss and stainless corrosion. Selection of PVC pipes requires closed laying.

In addition, fiber reinforced plastics (FRP) is also used in many pumping stations.

The size of pipe diameter directly affects the investment of pumping station and the efficiency of pumping station. The increase of pipe diameter and the decrease of pipeline resistance can reduce the power consumption, but the one-time investment of pipeline increases accordingly. On the contrary, the decrease of pipe diameter can reduce the investment of pipeline, but the resistance of pipeline increases and the annual power consumption increases. Therefore, how to determine the economic and reasonable pipe diameter, there should be an optimal plan, which is economic pipe diameter.

In the preliminary design stage, the economic pipe diameter can also be determined according to the economic velocity. The economic velocity of the outdoor pipe of the pumping station is generally 1. 5 – 2. 0m/s with a net head of less than 50m, and 2. 0 – 2. 5m/s with a net head of 50 – 100m.

For installation and safe operation of outdoor pipeline, some necessary accessories and equipment should be attached, such as gate valves, expansion joints, clapping doors, ventilation holes, etc. Pipeline accessories vary with the length of the pipeline, the height of the head and the laying mode. Which accessories should be installed depends on the specific conditions of the pumping station.

10. 3. 6   Supporting structure of outdoor pipeline

In order to fix the outdoor pipeline and maintain its stability, eliminate the vibration and displacement caused by normal operation and accident shutdown, the outdoor pipeline of pumping station must be provided with a stable supporting structure, which can be classified into two forms, buttress and anchor block.

The buttresses are mainly used for long pipe sections, and their functions are to bear the gravity of pipelines and water body and to reduce vibration. The section size of the buttress can be designed according to the structure (Fig. 10. 14 and Fig. 10. 15). Except near the expansion joints, other buttresses should be arranged with equal spacing. The spacing between the buttresses of steel tube can generally be 5 – 10m; the prestressed reinforced concrete pipeline should adopt a continuous pipe seat or two buttresses per section. The length of prefabricated reinforced concrete pipes is mostly 5m per section, and the connection type is adopted. Two buttresses should be set in each section, which are located at 1/4 and 3/4 of each section.

The anchor block is mainly used at the turning point of the pipeline, and also for the long pipe section on the slope. A anchor block is usually set at an interval of 80m to 100m. Its main function is to bear various forces at the turning point of the pipeline and offset the sliding force caused by gravity on the slope. Pipeline turning must be equipped with anchor blocks, and the spacing of anchor blocks on straight section of open pipe should not exceed 100m. The expansion joints shall be arranged between the two anchor blocks, and the expansion joints shall be arranged at the upper end. The section size of the anchor block can be determined by specific force analysis and structural calculation.

There are two types of anchor blocks: closed type and open type. The closed type places the pipeline inside the anchor block as shown in Fig. 10. 16 (a), and the open type places the pipeline on the anchor block surface as shown in Fig. 10. 16 (b) . Closed anchor blocks and pipes are fixed firmly, while open anchor blocks are convenient for inspection and maintenance. Large and medium-sized pumping stations have complex stress conditions at the turning point of the outdoor pipeline, and closed anchor blocks are commonly used.

10. 4   Outlet passage

The outlet passage is the connecting channel connecting the outlet of the guide vane of the pump and the outlet sump, which is mostly used in large and medium-sized pumping stations with lower head. In general, reinforced concrete cast-in-place is used in the outlet passage, which is integrated with the pump house in structure, and has the characteristics of short length, large change of section shape and small hydraulic loss. There are many kinds of outlet passages, such as siphon outlet passages, straight pipe outlet passages, oblique outlet passages, and so on.

10. 4. 1 Function and design requirements of outlet passage

The function of the outlet passage is to make the flow turn and diffuse better in the process of flowing into the outflow sump from the outlet of the guide vane of the pump and recover the kinetic energy to the maximum extent without the occurrence of separation of flow or vortex. The flow pattern and kinetic energy recovery in the outlet passage determine the hydraulic loss of the outlet passage. For low-head pumping station, hydraulic loss of outlet passage accounts for a large proportion of the total head of the pump, which has a more obvious impact on the energy performance of the pump device. Thus, the outlet passage is also an important part of the pump device.

The *Design code for pumping station* sets forth the following requirements for the design of outlet passages:

(1) The form of the outlet connected with the outlet of the guide vane of the pump shall be determined according to the requirements of the pump mechanism and the pumping station.

(2) The change of passage profile should be relatively uniform, and the equivalent diffusion angle should be 8° to 12°.

(3) The outlet velocity should not be greater than 1.5m/s (the flow velocity should not be greater than 2.0m/s when a flap valve is installed at the outlet).

(4) Appropriate cut-off mode should be adopted.

(5) Maintenance door groove should be set at the outlet of straight pipe.

(6) The construction should be convenient.

Among the above requirements, item (2) is to ensure uniform water flow diffusion and avoid flow separation or vortices; item (3) is to minimize the kinetic energy loss at the outlet of flow passage; item (4) indicates that attention should be paid to the mode of flow interruption. On this basis, the requirements of reducing civil engineering investment and construction convenience can be properly considered.

10. 4. 2　Typical outlet passage

1. Siphon outlet passage

The siphon outlet passage (Fig. 10. 17) is suitable for vertical or oblique low-head pumping stations with little change in water level of the outlet sump. Its main advantage is convenient and reliable operation. When the pump stops, it can destroy siphon and cut off flow through vacuum breaking valve. In addition, the siphon outlet passage is also convenient to cross the dike without affecting the safety of the flood dike. At present, hundreds of large pumping stations have adopted siphon outlet passage. The drawbacks of siphon outlet passage are: large engineering volume, difficult construction, and easy to cause unit vibration if not properly designed, etc.

The working principle of the siphon outlet passage is shown in Fig. 10. 18. Before the start of the pump, the part above the water surface in the siphon tube is filled with air. After the start of the pump, the water level in the outlet passage rises rapidly. At this time, the air in the flow passage is compressed. When the pressure P_c in the flow passage is higher than a certain value of atmospheric pressure P_a [Fig. 10. 18 (a)], the vacuum breaking valve at the top of the hump is opened, and part of the air in the flow passage will be discharged through the vacuum breaking valve [Fig. 10. 18 (b)]. When the water level raised by the pump exceeds the bottom of the hump, the flows across the hump, down the shell of pipe like an overflow weir, and also carries the remaining air in the pipeline out of the outlet of the flow passage. After all the air in the flow passage is discharged and the flow fills the full flow passage, the start-up process of the pump ends and enters the normal operation state [Fig. 10. 18 (c)].

Since the elevation of the hump top of the siphon outlet passage is higher than the maximum outflow level, after the full pipe flow is formed, there must be negative pressure near the hump of the flow passage. It can be seen that this outlet passage utilizes the siphon principle. The process of siphon is essentially the process of water filling the pipe section, air discharging outside the pipe and forming a certain vacuum at the hump during the start-up of the pump. When the pump stops normally or accidentally, the vacuum breaking valve can be opened in time, and the negative pressure at the hump can be used to make the air enter the top of the flow passage and destroy the vacuum in the flow passage [Fig. 10. 18 (d)], so as to achieve the purpose of cutting off the flow.

The reasonable design of the flow passage not only reduces the hydraulic loss of the flow passage, but also ensures strong entrainment capacity during the start-up process, so as to shorten the time of siphon formation as much as possible.

2. Straight pipe outlet passage

The straight pipe outlet passage (Fig. 10. 19) is connected with the pump outlet elbow. The section shape of the flow passage changes from circle to square. It enlarges gradually in both plane and elevation directions. Mean flow velocity in the flow passage decreases gradually. Any section in the flow passage has a certain positive pressure. The section of straight pipe flow passage is simple in shape and convenient in construction. Flapping gate and rapid-drop gate are used as cut-off measures and have been widely used in large and medium-sized pumping stations.

In order to avoid unnecessary energy loss, the outlet of the outlet passage should be submerged below the lowest operating water level of the outlet sump, and the minimum submerged depth of the upper edge of the outlet of the outlet passage should be 0. 3 – 0. 5m. According to the relative dimensions of the central elevation of the outlet section of the pump outlet elbow and the minimum water level of the outlet sump, the straight pipe outlet passage has three layout forms, i. e. ascending, flat and descending, as shown in Fig. 10. 20.

Straight pipe outlet passages should be equipped with ventilation holes. The purpose is to make the unit exhaust by the ventilation holes in the start-up stage and supplement air by the ventilation holes in the shutdown stage, so as to reduce the pressure pulsation in the flow passages.

The arrangement of ventilation holes depends on the layout form of the flow passage. For the descending flow passage or the curved low hump flow passage, the ventilation holes should be arranged at the highest position of the flow passage, and for the ascending outlet passage, they can be arranged near the outlet of the flow passage.

3. Oblique outlet passage

The oblique outlet passage is used in conjunction with the oblique shaft extension pump device. Its inlet is directly connected with the outlet of the guide vane of the pump. The angle α between the axis of the pump and the horizontal direction is usually 45°, 30° or 15°, as shown in Fig. 10. 21. The oblique outlet passage can be classified into two parts, bending section and straight section. The cross-section shape of the bending section varies from circular to square, the inlet section of the bending section is circular, and the outlet section is rectangular. The cross-section enlarges gradually in both the plane direction and the elevation direction, and the average flow velocity in the flow passage decreases gradually. Since the driving device of the inclined shaft extension pump device is located above the outlet passage, the bending section bends downward very badly in order to leave necessary space for the motor and its cooling duct. The smaller the angle α between the pump shaft and the horizontal direction, the greater the bending. The section shape of the straight section is rectangular, and the dimensions in the plane direction and the elevation direction are gradually expanded in a linear manner, and the average flow velocity in the flow passage is gradually reduced.

Like the straight pipe outlet passage, any section in the oblique outlet passage has a certain positive pressure, and the requirements in the cut-off mode and ventilation hole are the same as those in the straight pipe outlet passage.

10. 4. 3   Hydraulic design of outlet passages

1. One-dimensional hydraulic design method

Based on the one-dimensional flow theory, its main points can be summarized as follows:

(1) It is assumed that the average cross-sectional velocity is equal to the design discharge classified by the cross-sectional area.

(2) It is required that the average velocity of each section along the center line of the flow passage section be uniformly changed.

After the main dimensions of the outlet passage are preliminarily determined according to the empirical data, the profile of the longitudinal section of the outlet passage can be drawn according to the requirements of uniform change of the flow velocity, and then the plane profile can be drawn according to the method of decreasing the flow velocity. The specific steps are as follows:

(1) Firstly, a plane contour map is determined, and several sections are selected in the profile and plane contour map. The height of each section is obtained from the profile map, and the width of the section is obtained from the plane map, thus the area of each section is calculated.

(2) Calculate the average velocity of each section from the design discharge.

(3) The relationship curves of average velocity and length of flow passage, cross-sec-

tional area and length of flow passage are drawn. If the above two curves are smooth, they meet the requirements. Otherwise, the size of the section or the plan should be readjusted, and the steps of (1) to (3) should be repeated until the requirements are met.

2. Three-dimensional optimized hydraulic design method

The basic idea of three-dimensional optimized hydraulic design method for outlet passage is as follows: under the condition of given control size, given different outlet passage boundary, CFD calculation of corresponding flow field is completed, and three-dimensional flow state in the flow passage is investigated at different boundary. With the goal of no desquamation and vortices in the flow passage and minimal hydraulic loss in the flow passage, the geometric parameters of the flow passage are optimized one by one, and the profile of the flow passage is adjusted to gradually achieve the optimal hydraulic performance of the flow passage.

10. 4. 4   Cutoff mode

1. Flap valve

The flap valve is a kind of one-way valve, which is used together with straight pipe and oblique outlet passage. It is the most common way to cut off current. When the pump starts, the flap valve opens automatically under the impulse of flow; when the pump stops, the valve closes automatically depending on the weight of the flap valve and the force of reverse flow of water. The door top of the flap valve is linked with the door seat by hinges and the water stop between the door and the door seat is rubber. Its main features are simple structure, convenient application and low cost. It is widely used in pumping stations.

Normal flap valves, without any additional control equipment, are opened by the impulse of flow and closed by the force of self-weight or reverse flow, mostly for small and medium-sized pumping stations [Fig. 10. 22 (a)] . Because the flap valve opens under the impulse of flow, it often requires a certain amount of energy to be consumed during normal operation. In order to reduce the resistance of flapping, some large pumping stations use mechanical balanced flapping, floating box flapping, etc. [Fig. 10. 22 (b)] . On the other hand, since the door is closed by gravity or reverse flow, the door will have a higher velocity in the final closing phase, resulting in a greater impact. In order to avoid possible damage to the flaps and the door seats by high impact forces, some pumping stations use flaps with hydraulic buffers [Fig. 10. 22 (c)] .

2. Rapid-drop gate

The rapid-drop gate is usually equipped with a hydraulic hoist, which can be quickly opened and closed. It is another type of cut-off mode for large pumping stations, and is suitable for straight pipe and oblique outlet passages. The remarkable advantages of this cut-off mode are quick opening and closing and small impact force.

The "fast" of the rapid-drop gate is mainly to require the gate to fall quickly and cut off the flow when the pump stops. When the pump unit starts, if the gate is opened too slowly, for the axial flow pump unit, it will increase the starting head of the pump, which will lead to motor overload and unit vibration; if the gate is opened too fast, it may cause the water discharged from the pump and the flowing from the gate to collide in the flow passage, thus making it difficult to exhaust in the flow passage and causing greater pressure pulsation. In order to avoid problems caused by too fast or too slow gate opening during start-up of the unit, certain safety measures can be taken, such as setting breast wall or opening a small flap valve on the door page of the rapid-drop gate, as shown in Fig. 10. 23. After taking safety measures, the opening time and velocity of the rapid-drop gate need not be strictly required.

3. Vacuum breaking valve

The vacuum breaking valve is matched with the siphon outlet passage to destroy the vacuum at the top of the siphon channel and block the flow. After the shutdown of the pump unit, as long as the valve set at the top of the hump is opened and the air enters the runner, the vacuum can be destroyed and the flow can be cut off. The remarkable advantages of this cut-off mode are simple operation, reliable cut-off and convenient maintenance. In order to ensure the normal and safe operation of the unit, the vacuum breaking valve should meet the requirements of good sealing performance, rapid and reliable operation, and high exhaust sensitivity.

The most commonly used vacuum breaking valve is a pneumatic flat valve, which consists of valve seat, valve cover, cylinder, piston, piston rod, spring and other parts, as shown in Fig. 10. 24. If the vacuum breaking valve cannot be opened for some reason, the manual valve can also be opened, and compressed air can be fed into the cylinder of the vacuum breaking valve to make the vacuum breaking valve operate. If the vacuum breaking valve cannot be opened for special reasons, a large hammer can be used to break the plexiglass plate beside the vacuum damaging valve so that air enters the siphon pipe to ensure that the vacuum of the siphon pipe can be destroyed reliably and the water flow can be cut off after the pump is shut down.

(1) Hydraulic impulse vacuum breaking valve. The hydraulic impulse vacuum breaking valve is shown in Fig. 10. 25. When the pump is in normal operation, the flow flows along the arrow direction, and the baffle makes the baffle lever rotate around the fulcrum under the impact of the flow; thus, the valve plate compresses the air inlet to keep a certain degree of vacuum in the pipe. When the pump stops running, the flows backward and the baffle rotates reversely around the fulcrum by the backflow action, opening the air inlet, and the air enters the pipeline, destroying the vacuum, thus playing a role of shutoff.

(2) Ventilation pipe vacuum breaking valve. The way to destroy the vacuum is to guide the air inlet from the hump top to a certain position upstream below the hump top elevation, as shown in Fig. 10. 26. In this way, when the pump is in normal operation, the

water level rises due to the high upstream pressure and the intake port is submerged, thus playing a sealing role to facilitate the formation of vacuum. When the pump stops and the flow reverses, the water pressure in the original upstream decreases and the water level in the sump decreases accordingly. When the water level drops below the intake port, the air enters the pipeline and destroys the vacuum and cuts off the flow. The structure of this vacuum breaking valve is simple, but the variation of outlet water level can not be too large due to the restriction of outlet water level change.

参 考 文 献

［1］ 刘竹溪. 水泵及水泵站［M］. 北京：水利电力出版社，1986.

［2］ 田家山. 水泵及水泵站［M］. 上海：交通大学出版社，1989.

［3］ 刘超. 水泵及水泵站［M］. 北京：中国水利水电出版社，2009.

［4］ 武汉水利电力学院. 水泵及水泵站［M］. 北京：水利电力出版社，1981.

［5］ 沈日迈. 江都排灌站［M］. 北京：水利电力出版社，1985.

［6］ 姜乃昌. 水泵及水泵站［M］. 北京：中国建筑工业出版社，1998.

［7］ 刘超. 排灌机械［M］. 南京：河海大学出版社，2000.

［8］ 刘竹溪，冯广志. 中国泵站工程［M］. 北京：水利电力出版社，1993.

［9］ 湖北省水利勘测设计院. 大型电力排灌站［M］. 北京：水利电力出版社，1984.

［10］ 陆林广，张仁田. 泵站进水流道优化水力设计［M］. 北京：中国水利水电出版社，1997.

［11］ 斯捷潘诺夫. 离心泵和轴流泵［M］. 北京：机械工业出版社，1980.

［12］ 洛马金. 离心泵与轴流泵［M］. 北京：机械工业出版社，1978（1966）.

［13］ 日本农业土木专业协会. 泵站工程技术手册［M］. 丘传忻，等，译. 北京：中国农业出版社，1997.

［14］ 潘家铮，张泽祯. 中国北方地区水资源的合理配置和南水北调问题［M］. 北京：中国水利水电出版社，2001.

［15］ 水利部国际合作与科技司. 水利技术标准汇编（灌溉排水卷）［M］. 北京：中国水利水电出版社，2002.

［16］ Karassik，Igor J. Messina，Joseph P. Conper，et al. Pump handbook［M］. 3rd ed. McGraw-Hill Professional，2001.

［17］ 中华人民共和国水利部能源部. SL 26—1992 水利水电工程技术术语标准［S］. 北京：中国水利水电出版社，1992.

［18］ James B，Rishel PE. Water pumps and pumping systems［M］. McGraw-Hill Professional，2002.

术 语 表

Pump unit　水泵机组

Pump system　水泵装置

Pumping device　抽水装置

Characteristics　性能

System head curve　装置扬程曲线

Head curve of series pumping system　水泵串联扬程曲线

Head curve of parallel series pumping system　水泵并联扬程曲线

Requirement head curve　需要扬程曲线

Weighted average head　加权平均扬程

Variable vane angle governing　变角调节

Variable speed governing　变速调节

Pump duty (operating) point　水泵工况点

Pumping station　泵站（抽水站、扬水站，抽水机站）

Irrigation pumping station　灌溉泵站

Drainage pumping station　排涝泵站

Multistage pumping station　多级泵站

High-head pumping station　高扬程泵站

Low-head pumping station　低扬程泵站

Floating pumping station　低扬程泵站

Static head　几何扬程（静扬程，实际扬程，地形扬程）

Total head　总扬程

Installation height　安装高程

Pump house　主泵房（厂房）

Water retaining pump house　堤身式泵房

Pump house at levee-toe　堤后式泵房

Outdoor pump house　开敞式泵房

Submergible pump house　淹没式泵房（潜水式泵房）

Dry-pit pump house　干室型泵房

Wet-pit pump house　湿室型泵房

Block-foundation pump house　块基型泵房

Separated-foundation pump house　分基型泵房

Sliding pump carriage　泵车

Inlet passage　进水流道

Outlet passage　出水流道

Hydrocone　导水锥

Floor clearance　悬空高

Draft　吃水深度

Submergence　淹水深度

Suction sump　进水池

Outlet sump　出水池

Flap valve　拍门

Centrifugal pump　离心泵

Axial flow pump　轴流泵

Mixed flow pump　混流泵

Single-suction pump　单吸泵

Double-suction pump　双吸泵

Multi-stage pump　多级泵

Single-stage end-suction centrifugal pump　单级单吸离心泵

Single-stage double-suction centrifugal pump　单级双吸离心泵

Submersible motor pump　潜水电泵

Impeller　叶轮

Blade　叶片

Impeller seals　叶轮密封

Casing　泵室

Diffusion vane　导叶

Discharge　流量

Theoretical head　理论扬程

Output power of pump　水泵输出功率

Input power of pump　水泵输入功率

Pump efficiency　水泵效率

Hydraulic efficiency of pump　水泵水力效率

Volumetric efficiency of pump　水泵容积效率

Mechanical efficiency of pump　水泵机械效率

Static suction head　吸水高度

Net positive suction head　(NPSH) 汽蚀余量

Cavitation speed of pump　水泵比转速

Suction specific speed　汽蚀比转速

Head-discharge curve　流量-扬程曲线

Power-discharge curve　流量-功率曲线

Efficiency discharge curve　流量-效率曲线

Outlet pressure tank　压力水箱

Forebay　前池